KT-366-130

MOOD GENES

Hunting for Origins of Mania and Depression

Samuel H. Barondes

PENGUIN BOOKS

PENGUIN BOOKS

Published by the Penguin Group
Penguin Books Ltd, 27 Wrights Lane, London W8 5TZ, England
Penguin Putnam Inc., 375 Hudson Street, New York, New York 10014, USA
Penguin Books Australia Ltd, Ringwood, Victoria, Australia
Penguin Books Canada Ltd, 10 Alcorn Avenue, Toronto, Ontario, Canada M4V 3B2
Penguin Books (NZ) Ltd, Private Bag 102902, NSMC, Auckland, New Zealand

Penguin Books Ltd, Registered Offices: Harmondsworth, Middlesex, England

First published in the USA by W. H. Freeman and Company 1998
First published in Great Britain in Penguin Books 1999
1 3 5 7 9 10 8 6 4 2

Printed in England by Clays Ltd, St Ives plc

PENGUIN BOOKS

MOOD GENES

Samuel H. Barondes, MD, is the Jeanne and Sanford Robertson Professor and Director of the Center for Neurobiology and Psychiatry at the University of California, San Francisco. A member of the Institute of Medicine of the National Academy of Sciences, he also serves as President of the McKnight Endowment Fund for Neuroscience and recently chaired the Workgroup on Genetics of the National Institute of Mental Health. He is the author of Scientific American Library's *Molecules and Mental Illness*.

In Loving Memory of

Ellen Slater Barondes
Yetta Kaplow Barondes
Solomon Barondes

And to Welcome
Jonah Lewis Barondes Feingold

CONTENTS

PROLOGUE

To lighten the affliction of insanity by all human means is not to restore the greatest of the divine gifts; and those who devote themselves to the task do not pretend that it is. . . . Nevertheless, reader, if you can do a little in any good direction—do it. It will be much, some day.

—Charles Dickens (1852)

Every aspect of human behavior, from the conventional to the bizarre, reflects both personal experiences and inherited inclinations. Until recently, we mainly paid attention to the experiential component because its effects, being so obvious, seemed far more important.

But this is changing as scientists have learned to go beyond vaguely defined "inherited inclinations" to their very specific and tangible underpinnings—the genes themselves. Since the discovery of DNA, a powerful technology has been developed to scrutinize exact differences in the genes of individual people and to relate them to many attributes, such as vulnerabilities to various diseases. Now this same approach is being applied to hunt for DNA variations that play a role in "the affliction[s] of insanity"—afflictions that once seemed so heavily influenced by life experiences that genetic studies didn't seem very worthwhile.

Of these afflictions, one that is attracting particular attention is manic-depressive illness. Characterized by episodic and disruptive mood fluctuations, this illness is especially important because it affects so many people—about one in a hundred of us in its most flagrant form, and possibly several times as many of us in milder versions. Among those affected are many groupings of relatives, such as Charles Dickens and his father, John—who was sent to debtors' prison because of the uncontrollable spending so characteristic of mania. It is, in fact, the concentration of mood disorders in certain families—such as those discussed in this book—that first raised the possibility that genes are involved.

This book describes the hunt for the genes—which I will call "mood-disorder genes" or, for brevity, "mood genes"—that influence susceptibility to manic-depressive illness, and the tools that scientists are using to find these unique bits of DNA. What makes this hunt so noteworthy is that it does not depend on any preconceptions about the nature of mood genes. Starting with nothing more than a distinctive human behavioral pattern—manic-depression—and samples of DNA from a sufficient number of relatives who have this disorder, geneticists have devised ways to discover previously unknown genes that play a role in its development. They are, in fact, accustomed to beginning their work from a position of ignorance about the genes that control the biological processes they are examining, since many genes (and most functions of known genes) remain undiscovered. Because other mental illnesses, such as schizophrenia, autism, Tourette's syndrome, and Alzheimer's disease are also being studied in this way, it is likely that the basis of many forms of "the affliction of insanity" will be clarified as these gene hunts succeed.

The aim of each hunt is not simply to identify the genetic variations that may lead to the aberrant behavior, but to use this knowledge to find better ways to cope with it. Even in the case of manic-depression, for which there are some effective medications, such as lithium and Prozac, these drugs have many drawbacks. Treating a mood disorder with them is like treating an infection with aspirin: symptoms may be relieved but the fundamental problem remains unaddressed.

Finding mood genes will change all this. Their discovery will help us understand the underlying differences in brain chemistry of people who develop severe mood swings, and these differences will become new targets for biological and behavioral treatments. Eventually, their discovery will even teach us how to prevent the development of this terrible source of suffering that haunts many millions of people around the world and that poses an awful threat to their descendants.

But none of this will come easily. Though manic-depressive illness is an appealing target for a gene hunt (because genes seem to play such an important role in its development), it is also a challenging one (because multiple genes and environmental factors may be involved in an individual case). Also very challenging are the questions that are sure to be raised once mood genes are discovered. Should people be permitted (or encouraged—or even required) to have their mood genes examined to assess their chances of developing a mood disorder? Should there be prenatal testing for mood-gene variations that increase susceptibility to manic-depression? Answering these questions is particularly difficult not only because of the ethical issues they pose but also because it is likely that the same genetic variations that bring frantic mania and suicidal depression to some people may also enrich many aspects of life, as in the case of John and Charles Dickens.

Despite these challenges, finding mood genes will bring many benefits. Not the least of these will be a clearer definition of the relationship of normal mood fluctuations to the patterns that we define as mood disorders, and of the boundaries that we have erected to separate them. And even though there may prove to be quite a few mood genes, extending the hunt for many years, we may confidently predict, like Charles Dickens, that the consequences of identifying these tiny bits of DNA "will be much, some day."

1
MICHAEL'S FAMILY

That . . . inbred cause of Melancholy is our temperature [temperament], in whole or part, which we receive from our parents . . . it being a hereditary disease; . . . such as the temperature of the father is, such is the son's, and look what disease the father had when he begot him, his son will have after him. . . . And that which is more to be wondered at, it skips in some families the father, and goes to the son, or takes every other, and sometimes every third in a lineal descent, and doth not always produce the same, but some like, and a symbolizing disease.

—Robert Burton, *The Anatomy of Melancholy* (1621)

When Michael was in kindergarten his mother, Flora, was overwhelmed by her first attack of mania. Then twenty-four years old, Flora had just published her first novel to great critical acclaim. In light of this early success, her exuberance initially seemed understandable. But instead of coming down to earth, Flora kept getting more and more excited. At a party in her honor, she startled the guests by dancing on the table and stripping off her clothes.

A few restless, sleepless days later, this brilliant and elegant young woman had become so agitated and disheveled that she was admitted to a psychiatric hospital. Her diagnosis, confirmed by several specialists, was manic-depressive illness, a remarkably common mental disorder that affects about one in a hundred men and women. Because this was in 1941, before tranquilizers and lithium had been developed as treatments for manic behavior, it took more than a month of hospitalization and large doses

5

of sedatives to help calm her down. All that Michael remembers of this period is that he was very upset by her long absence.

In the years that followed, Flora was hospitalized many more times for flagrant manic episodes. Usually when they began she felt a growing sense of confidence and optimism, such as we may all experience when our lives seem to be going particularly well. But as the episodes developed, pleasant excitement was replaced by growing irritability, frequently accompanied by heavy drinking. Her mind began racing too quickly to be productive, and her work became garbled and useless. She often complained that people were stealing her ideas, that she was not properly valued by the critics, that she should have won a major literary prize. During one of these episodes she had a highly publicized affair with a well-known actor, which precipitated her divorce from Michael's father, Jacob.

Sometimes, after days of hyperactivity, Flora crashed into a deep depression—a common pattern in people with manic-depressive illness. Then she stayed in bed all day and paced all night, blamed herself for everything, stopped eating, looked despondent and unkempt. During one depressed period she told her psychiatrist that she was saving up sleeping pills to kill herself. Because antidepressants and mood-stabilizing drugs were still not available, shock treatments were given. After four or five of these her depression began to clear.

In addition to the serious breakdowns, Flora had dozens of others that didn't result in her admission to the hospital. Often they began for no apparent reason; but some, like her initial attack, were preceded by a significant professional success. One memorable episode, which made the gossip columns, began immediately after she had sold a two-part story to a major magazine. When the check came, she bought nine dresses, most of them tight and skimpy, all of them red. Every night she went out with friends, wearing a new red dress. She said this was her red period because her story was being read. She thought this pun was hilarious and she danced around, laughing, whenever she repeated it. She was high as a kite.

But red dresses were only one small part of this particular period of mania, which waxed and waned over several months. It occurred at a time when Flora's work was going particularly well. That summer, writing at least fifteen hours a day, she produced almost three hundred typed pages in three weeks, the core of one of her most successful novels.

Because of such bursts of creative energy, Flora considered mania—"my gemlike flame"—mostly a blessing. Although it was generally accompanied by profligate spending and sexual promiscuity that often brought her near to disaster, Flora was nevertheless very good at damage control. Michael was also a big help in smoothing things over. Beginning in his teens he had become an expert in soothing the ruffled feelings of those friends she offended while manic. They were willing to overlook her periods of grandiosity and irritability because when well she was so friendly and open and generous. Besides, she was a fascinating character—the life of the party—and such a brilliant writer.

I FIRST LEARNED about Michael's family in 1978, while attending a conference about new methods for studying brain function. Michael, who shared his mother's brilliance and eloquence, was one of a handful of molecular biologists and geneticists who were invited to spend a week teaching the rest of us, mainly neuroscientists, how to analyze DNA using the new techniques they were developing. To promote conversations between people from these different fields, the organizers of the conference had brought us together in an isolated mountain hotel and had selected pairs of us to share the small number of available rooms. Michael and I were assigned as roommates.

Once Michael found out that I was a psychiatrist it didn't take him long to tell me about the mental illness in his family. He was at the time particularly worried about Flora because she was back in the hospital, having precipitately stopped taking the lithium that she had, by then, been on for many years. But as he related her story it soon became clear that his concerns had been amplified because of his uncle Max.

Max, Flora's older brother and her only sibling, was in many ways her opposite: short, fair, and ungainly while she was tall, dark, and impeccably groomed; easygoing and the butt of his own jokes, while Flora tended to take herself more seriously. One of the few things they had in common was a facility with words, which in Max's case led him to become a writer of lyrics for popular songs and commercials.

Max also shared Flora's devotion to Michael. Unmarried and a loner, Max doted on his only nephew. When Jacob moved to another city after he and Flora were divorced, Max became Michael's surrogate father. In contrast to Flora, whose love for Michael was sometimes obscured by her unpredictability and whom Michael regarded from childhood as "my darling crazy mother," Max was always solid and reliable. Until the winter day in Michael's senior year in high school when Max, then thirty-eight years old, first tried to kill himself.

There was no apparent reason for Max's desperate action. The only unusual thing Michael could remember about the period leading up to it was that Max, who lived only blocks away and who usually dropped by almost daily, had not visited as much. Then, early one morning, Max went to the garage of his house, turned on his car's engine, and waited to die. Miraculously, a passing neighbor heard the car, found Max, and called an ambulance. Because Max was still threatening to kill himself after a week in the hospital, he was given several shock treatments, and his mood greatly improved. He also started psychotherapy, which he continued for many years, eventually going on to psychoanalysis five times a week, which he found very helpful.

But the insight he gained from this treatment didn't spare Max from more bouts of depression. Nor did a drug called imipramine—the grandparent of contemporary antidepressants such as Prozac—which Max began to take in the early 1960s, not long after its discovery. On two later occasions he again needed hospitalization, when self-loathing and the danger of suicide returned. But interspersed between bouts of depression were whole years when Max was quite normal. And he never became manic like his sister.

Yet the similarity between Max's and Flora's depressive episodes troubled Michael deeply. Even though Max's diagnosis was major depression, which lacks the manic episodes of Flora's illness (and affects at least five times as many people) several psychiatrists had told Michael that repeated attacks of such severe depression might really be an alternative form of the same illness Flora had. Michael's next questions were inescapable: Did the fact that two close relatives had serious mood disorders mean that others in his family would also be affected? What were the chances that he himself would eventually be stricken? Or his teenage children, Jerry and Charlotte?

Over the years Michael had learned that experts disagreed not only about the relationship between Max's and Flora's patterns of symptoms, but also about the origins of such illnesses. On one side was the view that prevailed through that period, the 1970s, that mood disorders were caused primarily by childhood traumas and parental styles. So finding that both Flora and Max had serious mood disorders was hardly surprising: their parents must have raised them both in a particular way that led to this outcome. Furthermore, proper child rearing should prevent the transmission of the problem to the next generation. But there were also a few psychiatrists who vehemently disagreed. To them the same finding meant instead that the shared mood disorders were mainly a reflection of shared genes. Though life events could certainly trigger mood disorders, having a particular genetic predisposition was, in their view, indispensable. And even though good upbringing might help prevent the development of such disorders, the vulnerability would be passed down with the genes.

To Michael both arguments seemed plausible. Already convinced of the importance of upbringing, he and his wife, Marcia, did their best to provide their children with what they believed was an ideal mixture of expectations and support. But would careful nurturing offer them any protection from a mood disorder that was primarily genetic? And if genes were really the main cause of Flora's and Max's disturbances, what were the chances that these same genes had been passed on—via Michael—to Jerry? Or Charlotte? Or both?

AT THE TIME I had this conversation with Michael, early in 1978, the main ideas about the inheritance of human diseases had not changed much since the start of the century. It was then that the British physician Archibald Garrod was doing his pioneering work on the inheritance of a rare form of arthritis called alkaptonuria (abbreviated AKU) which sometimes affected siblings—the first clue that AKU is hereditary. Although Michael and I only mentioned it in passing, our shared understanding of AKU played an important role in our discussion of his family.

People with AKU have one distinctive characteristic: brown urine. The color is imparted by a pigment, discovered in 1859, named alkapton. The prefix "alk" reflects the observation that formation of the pigment is accelerated.by alkali. So if the urine of a person with AKU is acidic (which urine generally is), it will have a fairly normal color. One test for AKU is to add alkali to the urine to accelerate formation of alkapton.

Although the genetic defect responsible for AKU is present at birth, the condition can go undetected. After all, slightly brownish urine is easily overlooked. Sometimes the tipoff is the intensification of stains in washed diapers, which comes about when alkaline soaps cause formation of alkapton. But bleach can obscure this chemical reaction. And there is no known harm to the infant if the diagnosis is missed.

Some harm does come from AKU, but not until middle age. By then a great deal of pigment has been deposited in body tissues, especially joint cartilage. The blackening of the joints is associated with the development of arthritis of the hips, shoulders, and spine. Pigment also accumulates at more visible sites. Middle-aged people with AKU may have gray ears and black patches in the whites of their eyes—not desirable features, but hardly life-threatening or disabling.

Awareness of the AKU story helped frame in two ways the issues that Michael and I discussed. First AKU is a good example of the insidious nature of many genetic disorders, which may not become manifest until late in life. Although often noticed in infancy because of the colored urine, AKU may escape detection or be dismissed as a minor oddity until arthritis develops around the age of fifty. In the same way, Flora's not showing obvious signs of a mood disorder until she was twenty-four and Max's not developing a serious depression until he was even older are consistent with the possibility that their symptoms are attributable to a genetic predisposition that had been latent until then.

The second intriguing feature of AKU is that its hereditary nature was heralded by brown urine—a very distinctive and easily measurable feature. Were there such an obvious tipoff to the nature of an inherited abnormality underlying manic-depression, no one would be arguing about whether or not genes played a role in this disorder. On the other hand, were alkapton colorless rather than brown, the very existence of AKU might still not have been discovered, let alone its genetic origin. There are, after all, millions of middle-aged people with mild arthritis of unknown cause, and many families in which more than one member is affected. So the AKU cases, a minute fraction of the total (AKU affects only a few people per million), might still be lumped together with these others, their cause still obscure. It is the distinctive pigment in the urine that was the essential factor in separating out AKU from other forms of arthritis; and it is the pigment that eventually pointed the way to the role of an abnormal gene in AKU.

Because urine has provided clues to the fundamental nature of many hereditary diseases, including common ones such as diabetes, there was a long period—from the 1940s through the 1980s—when psychiatrists collected it by the tankload. Even though unusual concentrations of chemicals in the urine could reflect external influences, such as dietary differences, still it seemed possible that people with mental disorders just might be excreting substances that would point to a cause. But despite enormous effort, no one has ever found an equivalent of alkapton for this or any other common mental disorder. The only good thing that came from the search for an abnormality in the urine of patients with manic-depressive illness was John Cade's accidental 1948 discovery (which I will return to later) that lithium is useful in the treatment of mania.

Had an "alkapton" been discovered for manic-depression, it would have been expected to lead eventually to an understanding of the biological basis of the disorder and thus to effective ways to prevent its development. Were this the case, Michael and his wife could decide whether or when they should have their children tested. If they did, and the results were positive, they could decide whether and when to institute the recommended preventative psychological or pharmacological measures. But in the absence of such a test, there was simply nothing to be done. All we could do was hope that if specific genes really played a part in Flora's and Max's mood disorders, they had not been transmitted to Michael's children.

A FEW YEARS after our meeting Michael called to tell me our hope was in vain. His son, Jerry, a college sophomore, had been hospitalized because of a pattern of behavior much like Flora's.

Jerry, then eighteen, had been everything his parents had wished for. A prodigious student in high school, he was also a star of the track team, outgoing and popular, with the dark good looks and confident bearing of both his father and his paternal grandmother. Accepted at an excellent college, Jerry quickly and successfully responded to its tough academic demands. In his sophomore year he decided to major in psychology and computer sciences, and met his first serious girlfriend, Lorraine. When Jerry came home for winter break he seemed happy and optimistic.

Soon after he returned to school, however, Jerry became obsessed with a research project on dating he had undertaken as part of a course in social psychology. He had received permission to create a questionnaire to study predictors of a successful first date and to administer the survey to several hundred members of his class. He would pair the respondents up on dates based on the answers and then assess the results. But when his closest friend, Bob, read the questionnaire and found that it mainly concerned personal sexual experiences, he urged Jerry to tone it down and prepare a revision.

The revision was worse, really bizarre. The original four-page questionnaire was now five times as long. Many of the inappropriate sexual questions had become flagrant obscenity. Alarmed, Bob showed it to Lorraine. When she agreed that it was completely out of line, they tried to persuade Jerry to drop the whole project before he got into serious trouble.

They had not bargained on Jerry's response. As they voiced their reservations, he became increasingly agitated and angry. They were pitifully conservative, he told them; the questionnaire would revolutionize dating services and would make him as rich and famous as *Playboy* had made Hugh Hefner. But even as he countered their arguments, Jerry acknowledged how tired and nervous he felt. Then he started to cry.

Convinced that Jerry needed professional help, his friends managed to get him to the student health service. Once there he was seen by two psychiatrists—a young resident from a nearby teaching hospital and Dr. R., a senior member of the faculty. With them, all he would talk about was the question-naire, explaining gleefully how he would become the national tsar of the dating business, then expand to women's lingerie. When they expressed skepticism he became angry. Pressed for details, he started hinting that many of the ideas had come to him from a female voice that he sometimes heard speaking to him in his head.

When Dr. R. called Michael, he said that Jerry had a serious problem and needed to be hospitalized. He was not sure of the diagnosis, but it was dangerous to let him return home.

DURING HIS FIRST meeting with Jerry, Dr. R. had conclud-ed that Jerry was probably suffering from manic-depressive ill-ness (or, in the new terminology of the time, "bipolar disorder"). For even though no single feature of his behavior was sufficient to clinch the matter, Jerry's combination of symptoms—elevated mood, grandiosity, decreased need for sleep, agitation, sexual preoccupation, irritability, and poor judgment—together pointed to this diagnosis. Furthermore, Jerry's wild conduct was not a minor behavioral aberration: it was serious and sustained enough to interfere with his schoolwork and his relationships.

Jerry's report of having heard voices was especially troublesome. Though not unusual in people with manic-depression, auditory hallucinations are particularly characteristic of another mental illness—schizophrenia—which is generally even more disabling. Like manic-depressive illness, schizophrenia frequently strikes in late adolescence or early adulthood, so this possibility had to be taken very seriously. But Jerry's pattern of behavior, particularly his clear signs of elevated mood, were in marked contrast to the detachment and withdrawal commonly seen in schizophrenia. As Dr. R. went through a formal checklist of symptoms, manic-depressive illness fit best.

Before settling on this diagnosis, Dr. R. felt that other possibilities had to be excluded. Jerry's urine was checked for signs of drug use, since cocaine, amphetamine, and other drugs may produce a similar pattern of behavior. He was also tested for a number of general medical conditions, such as hyperactivity of the thyroid gland, the effects of which can sometimes be mistaken for manic-depressive illness. When these tests were all found to be negative, Dr. R. was convinced. Further confirmation came when Michael told him about Flora and Max. Manic-depressive illness and schizophrenia tend to run in different families.

Although the diagnosis of manic-depressive illness implies lifelong vulnerability to attacks, there was also good news. Jerry recovered very quickly. Within two weeks his behavior seemed normal. Jerry had been started on lithium pills as soon as all the medical tests were done, but improvement was evident even before the effect of the drug was to be expected. Instead of leading Dr. R. to reevaluate the diagnosis, Jerry's rapid recovery served as confirmatory evidence. Rapid and complete recovery from manic episodes is, fortunately, common—even without medication.

Because of this quick improvement, Jerry was back attending classes within ten days of his hospitalization. For the first few nights he continued to sleep in the hospital, but a week later he had returned to his apartment. Nevertheless, Dr. R. continued the treatment with lithium, because it not only relieves acute manic episodes but also prevents new attacks. It can also be effective in preventing episodes of severe depression that Jerry, like Flora, would likely be prone to in the future. This tendency to swing between the poles of mania and depression is, indeed, the basis for the diagnostic terms "manic-depressive illness" and "bipolar disorder."

IN THE YEAR after Jerry's manic episode, Michael and I had many conversations about the nature of the mental illness in his family. For him it was useful to have found a colleague who knew something about the affliction that constantly threatened his mother, his uncle, and his son. For me, these talks began to take on a professional significance. I had treated many patients with serious mood disorders and hoped that new tools would eventually be developed with which to investigate their biological roots. My conversations with Michael about new ways of examining human DNA helped me realize that the time was near when that hope might be fulfilled.

This hope had first been kindled when I was an undergraduate at Columbia in the early 1950s by a teacher, Murray Jarvik, whose research on LSD led me to the startling realization that simple chemicals can control complicated mental processes. And though it had begun to fizzle when I moved on to Columbia's medical school (where I learned how little was then known about the brain), it was reignited by my training with Gordon Tomkins at the National Institutes of Health (NIH) in Bethesda, Maryland, which I began in 1960—just in time to participate in the first flowering of the new field of molecular biology.

So new, in fact, that it had only been seven years earlier, in 1953, that Francis Crick and James Watson had established the structure of deoxyribonucleic acid, or DNA—the genetic material. Starting with the knowledge that DNA is made up of long strands of four simple chemical building blocks called bases (whose names—adenine, guanine, cytosine, and thymine—are abbreviated A, G, C, and T), Watson and Crick had the brilliant insight that they come together as pairs (called base pairs)—T pairing with A and C pairing with G—resulting in two perfectly paired strands, the famous double helix. And from that monumental contribution there soon developed what I will call (adopting a phrase first used in 1957 by Crick) the Central Dogma, which has guided biology ever since.

The crux of the Central Dogma is that the structure of a DNA molecule encodes information that can be used by the body only after it has been translated into another type of molecule called a protein. To this end the huge chains of DNA (which are, in many cases, more than one hundred million base pairs long) are divided into segments (usually tens of thousands of base pairs long), each of which constitutes a unit of information—a gene. The function of each gene is to send out a molecular "messenger," called messenger RNA, that eventually determines the structure and the ongoing manufacture of a particular protein, a working element of the body that controls a particular function (such as a step in the digestion of food).

At the time I started at the NIH, Gordon—himself a physician-scientist—was one of a small number of people who were thinking about the long-range medical implications of the Central Dogma. Brilliant, imaginative, and infectiously enthusiastic, he quickly persuaded me that this new approach would, as it developed, make possible a direct attack on all sorts of seemingly inscrutable biological processes, including those involved in mental illness. For just as the functioning of the intestines to digest food depends on the right balance of specific intestinal proteins (each controlled by a specific gene), so too does the functioning of the brain to process ideas and feelings depend on the right balance of specific brain proteins (each controlled by a specific gene). And just as an abnormality in the amount or composition of a single intestinal protein (such as a deficiency in a protein called lactase) can lead to an intestinal illness (lactose intolerance), so might an abnormality in the amount or composition of a brain protein (such as a protein involved in the manufacture of serotonin, a vital brain chemical) lead to alterations in mental activity that we call mental illness.

Before Gordon explained the Central Dogma, I had no idea that the revolution in biology that was going on around me would be applicable to mental disorders; afterward, its relevance to the control of brain chemistry made its potential applicability seem obvious. While Gordon recognized that we were, at the time, a long way from explaining the details of brain functioning in terms of the actions of the tens of thousands of different proteins which it makes use of, he was convinced that the Central Dogma had defined a path of systematic and cumulative research that would, step by step, lead us to that goal. So great, in his view, was the power of this revolution that, no matter how hard the problem, "anything is possible."

While at the NIH, I also had the good fortune to live through a dazzling verification of Gordon's optimism. This came by way of an apprenticeship (arranged by Gordon) with another young scientist, Marshall Nirenberg. As shy as Gordon was outgoing, and as sharply focused as Gordon was broad, Marshall was working on what seemed at the time a hopelessly difficult problem—the nature of the code used to translate the information in the arrangement of the four bases in a particular messenger RNA into the arrangement of the twenty amino acids that are used to make a particular protein. But by making use of a battery of different artificial messenger RNAs, each manufactured in a test tube, this genetic code was—to everyone's amazement—completely deciphered in Nirenberg's laboratory in just a few years. In a stunning series of experiments it was shown that a specific series of three bases (for example, GAA or GGC)—called a codon—is the instruction for the addition of a particular amino acid (for example, glutamic acid or glycine, respectively) at a particular place in the growing string of amino acids that, when completed, constitutes a protein. Even "molecular punctuation"—where to begin and end the translation of a string of bases—was eventually attributed to specific codons.

Having been at the right place at the right time, I therefore had the heady experience of being catapulted from total naïveté into charter membership in what the newspapers would soon call "The Code of Life Team." In fact, Marshall's work proved so successful that in 1968 he was on his way to Stockholm to receive a Nobel prize. And the genetic code that he deciphered now occupies the same honored place on the inside cover of many textbooks of biology that the periodic table of the elements occupies in many textbooks of chemistry. Although deciphering the code is only one step on the long road to understanding the molecular basis of biological functions—including those at work in the brain—it was indispensable for all subsequent research.

Convinced by this exciting series of events that, as Gordon had assured me, anything was really possible, I went on to three years of clinical training in psychiatry, in the hope of identifying an aspect of mental illness that I could address with these seemingly omnipotent tools. Then, coming down to earth, and recognizing that, for the moment, we were still too many steps away from being able to address the molecular aspects of these complicated disorders, I decided to bide my time by sticking to more accessible aspects of brain biology. Which brought me to the conference where I first met Michael.

At the time of the conference, late in 1978, NIH's investment in training physicians in biomedical science had begun to bear fruit in many fields, making even difficult problems such as the biological basis of mental illnesses increasingly accessible. Then, shortly after Jerry's hospitalization, the discovery of new tools for chopping up the strands of human DNA led to a conceptual breakthrough: a revolutionary new method for developing a map of all human genes. With such a map the vast uncharted regions of human DNA could be systematically examined in greater and greater detail, making possible the hunt for specific genetic variations that play a role in a particular inherited disease. As Michael and I continued our discussions, we became increasingly excited about the possibility that this approach might be applicable in elucidating and eventually alleviating the manic-depressive illness in his family.

Assuming, of course, that genes played an important role in mood disorders such as manic-depression and major depression, a hypothesis suggested but far from proved by many family histories, including Michael's. Stimulated by our conversations, and interested to explore further this assumption within his own family, Michael became curious about the possibility that other relatives, such as Flora's parents, might also have shown some signs of this disease.

WHILE STILL A TEENAGER, Flora had also been curious about her family history. She questioned both her parents, David and Miriam, about the small villages in Eastern Europe where they had grown up, seeking to know as much as she could pry out of them about their lives. Through the years she made good use of the material they provided in her stories and novels.

Miriam and David had both immigrated to America in their late teens, meeting on New York's Lower East Side and marrying in their mid-twenties. Both had worked in clothing stores, mostly in sales. David had been extremely successful, a supersalesman. Miriam was quieter, accurate and methodical but not inspiring. As Michael recalled what he had learned about his grandparents from Flora, he zeroed in on David as the more likely candidate for a mood disorder. In his opinion Miriam seemed too solid and even. David was more quixotic, the one who made their lives exciting.

One story about David made Michael especially suspicious. In the 1930s, David, by then a legendary dress salesman, decided to go into business for himself. His idea, revolutionary at the time, was to create a self-service dress store. Of course, David knew full well that selling dresses frequently required an attentive salesperson. But he also understood that many customers liked to be left alone and that, particularly at the low end of the business, there could be huge cost savings if the sales staff were kept to a minimum.

Despite Miriam's warnings that the idea would not work, David borrowed as much as he could. He rented a store in a popular neighborhood, worked day and night—and failed miserably. It was only because he was such a superb salesman that he could eventually get his former job back and pay off his debts. For more than a month after this adventure he stayed home mostly in bed, constantly berating himself. But he got over it.

Another reason Michael began to suspect that David had a mood disorder is that he learned from Flora that David was constantly chasing women. Though not all infidelity is grounds for a psychiatric diagnosis, unusually elevated moods often bring poor judgment in such matters. At one point David's behavior was so outrageous that Miriam considered a divorce. But things were patched up, and David allegedly reformed. Flora said she really had her doubts, that her mother had probably suffered in silence.

When Michael recollected these stories about David he thought back on his own memories and impressions of his grandfather. A bit of a dandy, with black hair meticulously pomaded and parted down the middle, nice suits carefully pressed. Also a jokester, playing little tricks, very energetic and upbeat, blue eyes sparkling. But in the end Michael wasn't really sure what he could say about his grandparents. From what he knew, Miriam was probably "normal," whereas David's mood seemed to fluctuate to extremes. Which posed some questions that Michael really couldn't answer: Does being jolly, risk-taking, and unfaithful mean that you have a gene that predisposes you to manic-depressive illness? Does the fact that David became depressed when his store went out of business indicate that he actually had a mild form of mood disorder? Michael recognized that David's behavior might well be viewed as a normal set of responses to his circumstances. He also recognized that the boundary between what was generally considered "normal" and "abnormal" was probably arbitrary and certainly difficult to define. It was only because of the flagrant mood disturbances in several of David's descendants—which Michael was quite willing to consider "abnormal"—that there was rea-

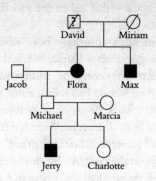

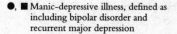

●, ■ Manic–depressive illness, defined as
including bipolar disorder and
recurrent major depression

son to wonder if David too was showing some suspicious signs
of a similar problem.

To summarize what he knew about mood disorders in his
family, Michael sent me a sketch of his family tree, drawn in the
form of a pedigree that geneticists use to describe patterns of
inheritance. Circles are symbols for females, squares for males;
the slashes indicate people who have died; filled–in symbols
indicate that a person is affected with the characteristic that is
under consideration—in this case a serious mood disorder. The
pedigree that Michael sent me appears above.

Michael had filled in Flora's and Jerry's symbols because they
both had unambiguous diagnoses of manic–depressive illness,
whereas symbols for unaffected members of his family, such as
that for his teenage daughter, Charlotte, were kept open. Max's
symbol was filled in because Michael had been persuaded by the
view of those experts who regarded serious and recurrent
depression like his uncle's as an alternative form of manic–
depressive illness. But the main point of this sketch was the question mark that Michael had drawn on David's symbol. It was
Michael's way of wondering if David's behavior was sufficiently

unusual to implicate him as the carrier of the genetic abnormality—if such it was—that led to the mood disorder in some of his descendants.

This question might also have been raised about Michael himself. Like his grandfather David, Michael was an unusually upbeat person, extremely energetic, enthusiastic, productive, and good humored. And Marcia had told me that when he encountered difficulties at work he would stay home for a few days, sleep poorly, and look distraught. Were I willing to accept the evidence Michael presented about his grandfather, I could easily justify a change in the pedigree by amending his own symbol as he had David's. As we shall see, I later had reason to remember this missing question mark.

IN THE TWO decades since my first meeting with Michael, human genetics has been transformed from an esoteric discipline concerned with rare diseases such as AKU to a front-page science that addresses the role of genes in fashioning a great many human attributes—including a great many common diseases such as heart attacks and cancer. What is making this possible is the development of powerful new ways to examine our genetic material—our DNA—and the concurrent development of computer-based storage and retrieval systems to help use this vast amount of information. With this new technology the precise DNA variations that account for inherited vulnerabilities are being identified with a rapidity that continues to amaze even those of us who work in this field.

As Michael and I had hoped, manic-depressive illness is among the disorders that are being studied in this way. Like the hunt for the genes that influence other human vulnerabilities, the hunt for "mood genes" depends on the accumulation of DNA samples from families in order to determine which genetic variations are shared by relatives who have the mood disorder and not by relatives who have been spared. But simply showing that a few affected relatives—such as Flora and Jerry and Max— all share a certain genetic variation is hardly sufficient for this purpose. To identify with confidence the genetic variations that are consistently shared by affected people and not by their unaffected relatives requires the cooperation of many families each with multiple members who have manic-depressive illness. Only by comparing DNA samples from large numbers of affected people with those from their unaffected relatives is it possible to establish that a particular difference in the DNA—that is, a particular genetic variation—is truly associated with manic-depressive illness because the variation is consistently passed down a pedigree along with the disease.

To establish this association it is also necessary to distinguish clearly those relatives who have the disease from those who are unaffected. Without being able to make this distinction reliably, some people who are not truly affected might be lumped together with some who are, and vice versa. In that event a hunt for a consistent genetic variation that distinguishes these two groups would be doomed to failure. For example, were Max's depression not really an alternative form of Flora's and Jerry's illness, he would not be expected to share the same mood gene variation as they do; and lumping all three of these relatives together in the same category would confuse the whole picture.

This is why the lack of an equivalent of alkapton for manic-depression is so regrettable. In its absence the hunt for mood genes must rely on a diagnostic scheme based solely on patterns of behavior. Unfortunately it is not easy to draw a sharp line between manic-depression and other mood variations. In fact, as we shall now see, the nature of this boundary has been a matter of heated controversy for about a hundred years.

2

A SINGLE MORBID
PROCESS?

In the course of the years I have become more and more convinced that . . . [circular insanity, simple mania, melancholia, and slight colourings of mood] only represent manifestations of a single morbid process.

—Emil Kraepelin (1915)

Even in descriptive psychiatry the definition of melancholia is uncertain; it takes on various clinical forms (some of them suggesting somatic rather than psychogenic affections) that do not seem definitely to warrant reduction to a unity.

—Sigmund Freud (1917)

Throughout the twentieth century, attempts to understand manic-depressive illness have themselves swung between poles established by the two great founders of modern psychiatry. Born in the same year, 1856, trained in medicine at a time when it was just beginning to establish itself on a scientific footing, they have each left a rich legacy. So great was their influence that both are included in a recent book on the hundred most influential scientists: one of them, Emil Kraepelin, ranked 92 (of *all* scientists, *ever*) and the other, Sigmund Freud, a stratospheric 6 (Isaac Newton is ranked 1, and Charles Darwin 4).

But despite their shared eminence, these two pioneers had very different views of manic-depression. Freud was mainly interested in it as an example of a particularly intense battle of internal psychological forces, and paid little attention to the evidence that it might reflect an inherited "somatic affection." In contrast, Emil Kraepelin was mainly interested in the importance of heredity in the development of not only manic-depressive illness but also "slight colourings of mood." And even though we now know that Kraepelin's view that all mood disorders are reflections of "a single morbid process" was an oversimplification, it is his emphasis on the hereditary nature of manic-depression that guides current research.

BORN IN NEUSTERLITZ, Germany, a rustic village near the Baltic, the son of an actor, Emil Kraepelin was quickly caught up in the effort to transform medicine from an art to a science. The aim enunciated by the great German pathologist Rudolf Virchow, one of Kraepelin's heroes (his other was Bismarck, the unifier of Germany), was to organize the bewildering array of human afflictions into well-defined entities by classifying them on the basis of symptoms and signs of illness coupled with pathological changes found on postmortem examination. In this way apparently similar diseases would be distinguished from each other, an indispensable step in the identification of their causes. A memorable example comes from an autopsy Virchow performed in 1866 on a sixty-seven-year-old man. In this case he discovered a new type of arthritis associated with black pigmentation of the joints that Archibald Garrod would subsequently show to be caused by AKU.

Kraepelin's interest in applying this same approach to psychiatry was already apparent in his medical school thesis, "On the Influence of Acute Diseases on the Origin of Mental Diseases," which he wrote while at the University of Wurzberg, where Virchow had taught. To supplement his training, which had been mainly restricted to medicine, Kraepelin subsequently moved to Leipzig, where he studied with Wilhelm Wundt, who had helped transform psychology from its origins in philosophy into a laboratory science. He also worked as an assistant to several distinguished psychiatrists and pathologists as he prepared for a faculty position at a university.

By the time he was thirty, in 1886, Kraepelin struck out on his own, becoming a professor of psychiatry at the University of Dorpat (now Tartu University in Estonia). Calling his inaugural lecture as a professor "The Directions of Psychiatric Research," Kraepelin outlined his Virchowian goal: to organize mental disorders into well-defined categories as the route to defining their causes. In 1883 he had already begun to publish a classification of mental disorders in the form of a textbook that he continued to revise through nine editions and that was, for many years, the definitive work in the field. The last edition, almost 2500 pages long, appeared in 1927, a year after his death.

The growing size of successive editions of Kraepelin's textbook reflected the rapid accumulation of knowledge about many forms of mental illness. Particularly dramatic was the series of discoveries about a prevalent mental illness of the time known as general paresis (or general paralysis of the insane), which often began with disturbances of mood, including periods of mania, and culminated in paralysis. Although the disease was long known to be accompanied by degeneration of the nervous system, the detailed microscopic changes in the brain were first meticulously described in 1904 by a then obscure physician, Alois Alzheimer, whom Kraepelin had just recruited to his new department in Munich. Two years later the cause of these degenerative changes was shown to be the microbe responsible for the venereal disease syphilis. And by 1910 Paul Ehrlich, a founder of pharmacology, had created an arsenic-containing drug, arsphenamine (soon called "Dr. Ehrlich's magic bullet"), which killed the microbe before it invaded the brain.

Not all cases of brain degeneration proved so easy to understand or treat, however. A notable example was discovered in 1905, by the same Alois Alzheimer, in the course of a microscopic examination of the shrunken brain of a fifty-one-year-old woman who had suffered from severe and progressive memory loss. The brain of this patient had two distinctive features: it was filled with microscopic deposits of a substance called amyloid; and many of the nerve cells contained tangled fibers. But unlike the degenerative changes caused by syphilis, the abnormalities Alzheimer saw had no obvious source. In fact, it is only through the recent identification of several genes implicated in Alzheimer's disease—the name Kraepelin gave it in honor of its discoverer—that scientists are beginning to understand how this degeneration comes about.

Yet as complicated as the story of Alzheimer's disease is turning out to be, at least there were specific pathological changes in the brain that helped to define it as a distinct entity. For many of Kraepelin's patients, not even Alzheimer's meticulous autopsies turned up signs of a brain abnormality. Nevertheless, Kraepelin believed that these patients also suffered from brain diseases. But how could he prove this? And what criteria could he use to categorize them?

Following the work of others, Kraepelin separated these patients into two broad categories on the basis of their symptoms. One category he called dementia praecox—"premature" dementia, the dementia of young people—now renamed schizophrenia. Patients in this category had prominent delusions, such as the belief that their minds were being controlled by outside forces or that they were the victims of organized plots; hallucinations, such as voices that gave them warnings or directed their actions; and a tendency to become progressively withdrawn and uncommunicative. Then, in the sixth edition of his textbook in 1899, Kraepelin brought together in a second umbrella category another large group of patients. Their unifying feature was prominent mood swings, which contrasted strikingly with the flattened emotions and withdrawal of dementia praecox. For these patients he invented a new diagnostic term: manic-depressive insanity.

Kraepelin's separation of dementia praecox from manic-depressive insanity, a distinction taken for granted today, was not a trivial achievement, because there is so much overlap in symptoms. Both may strike in adolescence or early adulthood, and patients in both categories may display bizarre behavior and suffer from delusions and hallucinations. In fact, there is still debate about the exact dividing line, a debate that will not be settled until we learn more about what causes these disorders. Not knowing the causes, Kraepelin emphasized the different life courses of typical patients from each category. Patients with dementia praecox showed progressive mental deterioration, moving further and further into a private inner world of delusions. In contrast, patients with manic-depressive insanity had prolonged periods of normality, and sometimes exceptional creativity, between their attacks; and they did not deteriorate mentally over time.

As Kraepelin gained experience with manic-depressive insanity he enlarged the category. From the start he included not only patients who had episodes of both mania and depression but others who (like Max) had only one or the other, thereby establishing a tradition of lumping all these conditions together that was maintained for many years. Eventually he also included people with milder mood disturbances. The broadening of this category was announced in the eighth edition of his textbook, which appeared in 1915:

> Manic-depressive insanity . . . includes on the one hand the whole domain of so-called *periodic and circular insanity*, on the other hand *simple mania*, the greater part of morbid states termed *melancholia* and also a not inconsiderable number of cases of *amentia* [intellectual impairment]. Lastly, we include here certain slight and slightest colourings of *mood*, some of them periodic, some of them continuously morbid, which on the one hand are to be regarded as the rudiment of more severe disorders, on the other hand pass over without sharp boundary into the domain of *personal predisposition*. In the course of the years I have become more and more convinced that all the above-mentioned states only represent manifestations of a *single morbid process*. [italics in original]

Kraepelin's reason for lumping together different forms of mood disorders—including those in which wild attacks of mania are the outstanding feature and those in which there is only stark depression—was that both forms are often found in the same families (as with Flora and Max), implying that they share the same underlying cause. To emphasize his interest in the familial nature of mood disorders, Kraepelin included in the eighth edition of his textbook a family's annotated pedigree (reproduced on the facing page), with males and females designated, respectively, ♂ and ♀. His study of this family indicated in his own words, "that of ten children of the same parents, who probably were both manic-depressive by predisposition, no fewer than seven fell ill in the same way; of the five descendants of the second generation four have already fallen ill."

From pedigrees of this sort Kraepelin concluded that the cause of manic-depressive insanity was a "hereditary taint," and that the same hereditary factors that led some to swing between mania and depression might lead their relatives only to depression. To pursue this proposal, Kraepelin established a Genealogical and Demographic Department, headed by Ernst Rüdin, in the Kaiser Wilhelm Institute for Psychiatry (the world's first institute for psychiatric research, now the Max Planck Institute for Psychiatry), which he founded in Munich in 1917—just as he had at the turn of the century established the laboratory of anatomy where Alzheimer did his famous work. The aim of Rüdin's department was to study the distribution of serious mental illnesses in the overall population and in particular families. Much of the early evidence for the familial nature of mental illnesses was accumulated at this institute, which Rüdin took over after Kraepelin's death.

But unlike his hiring of Alzheimer, which brought Kraepelin much reflected glory in his lifetime, the appointment of Rüdin would posthumously haunt his reputation. For Rüdin was not only convinced that mental illnesses were hereditary but also that they should be eradicated by preventing those affected from having children. When the Nazis came to power Rüdin was instrumental in drafting the infamous Law to Prevent Hereditarily Sick Offspring. Enacted in July 1933, it led to the compulsory sterilization of hundreds of thousands of people with "congenital mental defect, schizophrenia, manic-depressive psychosis, hereditary epilepsy, hereditary chorea, hereditary blindness, hereditary

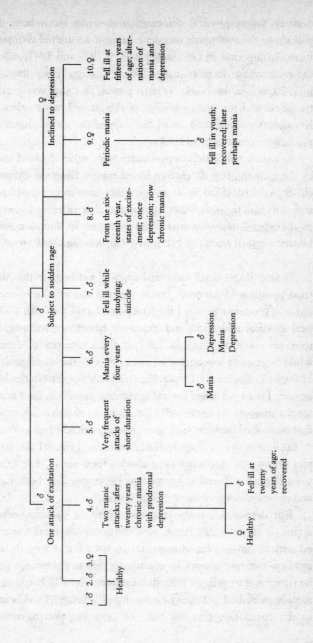

deafness, severe physical deformity and severe alcoholism"—a prelude to the systematic murder of almost a hundred thousand mentally ill people in Germany between 1939 and 1941, and to the even greater horrors that followed. Though Kraepelin died in 1926, before the Nazis came to power, his name was greatly tarnished; and the Nazi activities of Rüdin and many other of Kraepelin's students and associates stopped the development of psychiatric genetics in its tracks.

Kraepelin's approach to psychiatry was further shunted aside by the growth of an alternative. In contrast to Kraepelinian psychiatry, which, sullied by its associations, came to be considered authoritarian, rigid, and mechanical, the alternative was viewed as thoughtful, compassionate, and humane. Its founder, who himself escaped from the Nazis in 1938, was Sigmund Freud.

BORN IN THE small industrial town of Freiberg, in the Austrian province of Moravia, Freud was the son of a wool merchant of moderate means. Like Kraepelin, Freud had both a classical medical education and extensive laboratory training; he took his medical degree, in 1881, at the University of Vienna while beginning a program of research with the distinguished biologist Ernst Brücke. Displaying considerable talent in the laboratory, Freud published several significant papers on the microscopic anatomy of nerve cell clusters in a primitive fish, work that he looked back on with great nostalgia almost half a century later, writing to a disciple, Karl Abraham, "I think I was happier about that discovery than about others since." For many years thereafter Freud continued to be engaged in both basic and clinical studies of the nervous system.

But despite a twenty-year investment in the biomedical approach to psychiatry, Freud, now thirty-nine, became frustrated with its obvious limitations. Having tried for so long to find out how our minds work by examining cells in the brain, a goal he continued to struggle with in his notebooks of 1895 (posthumously published as *Project for a Scientific Psychology*), Freud came to the conclusion that the task was hopeless. Peering into a microscope wouldn't really tell him anything about the thoughts and feelings of his patients. To study such matters his consulting room would have to become his laboratory.

His specialty became a disorder called hysteria. The sufferers, young women, often teenagers, had multiple complaints—temporary paralysis, attacks of blindness, tingling sensations, fainting spells—that suggested an incurable degenerative disease of the brain. But Freud suspected that hysteria was neither incurable nor due to brain degeneration. Instead he believed that the source of his patients' symptoms could be found in their early sexual development; and that their symptoms could be relieved by the recollection and interpretation of such experiences through a form of treatment he called psychoanalysis. As he probed for details about their fantasies and their dreams, he became increasingly convinced that early childhood development and reactions to awakening sexuality were the major formative experiences of everyone's life. When his patients improved, he took that improvement as validation of both his treatment and his theories.

So too did many others who joined him in the practice of psychoanalysis. And it did not take long for them to extend Freud's approach to the more serious mental disorders that Kraepelin was mainly concerned with—schizophrenia and manic-depressive illness. Although Freud himself had only a limited interest in applying analytic techniques to these conditions because they seemed much too challenging, zealous followers soon began treating them with psychoanalysis.

GIVEN THIS HISTORY, it came as no surprise that my clinical training in psychiatry, which I started in 1963 at McLean Hospital (a Harvard teaching hospital in the Boston suburb of Belmont), would be based on the psychoanalytic view. Even though McLean specialized in the treatment of people with serious mental disorders—which is why I had chosen to train there—psychoanalysis was the main theoretical framework. In fact, a centerpiece of my education would be to conduct a form of psychotherapy based on psychoanalytic principles (but face to face instead of "on the couch" as in traditional psychoanalysis, and much more interactive) with a sixteen-year-old girl and a fifteen-year-old boy, both of whom had schizophrenia. Meeting with each of these seriously ill young people three times a week for two years, I struggled to help them cope with the periodic intensification of their symptoms, while trying to find out what the symptoms themselves might symbolize.

Attempts to understand the meaning of symptoms were also an important part of the treatment of the many people with manic-depressive illness who were hospitalized at McLean. Instead of something by Kraepelin, its foremost student, the main assigned reading on this disorder was *Mourning and Melancholia,* a 1917 essay by Freud and such a favorite of our training director, Alan Stone, that he also made it a central topic of his evening seminars. Accompanied by elegant Viennese-style pastries and top-grade cigars, these gatherings were held at his splendid house on Brattle Street in Cambridge, and Freud was always the center of attention. While I remained immersed in biological research throughout my psychiatric training, it was hard to resist the appeal of the wide-ranging conversations about human psychology that Alan Stone presided over in such a gemütlich environment.

One notable feature of *Mourning and Melancholia* is its exploration of a favorite interest of Freud's: the relationship between a normal mental process (in this case, grief in response to the loss of a loved one) and an abnormal one (in this case, a major depression, which may feel and appear exactly the same but is usually more sustained and without an obvious provocation). Another is its claim that psychoanalysis could be useful in explaining and treating not only depression but also mania:

> The most remarkable peculiarity of melancholia, and one most in need of explanation, is the tendency it displays to turn into mania accompanied by a completely opposite symptomatology. Not every melancholia has this fate, as we know. Many run their course in intermittent periods, in the

intervals of which signs of mania may be entirely absent or only very slight. Others show that regular alternation of melancholic and manic phases which has been classified as circular insanity. *One would be tempted to exclude these cases from among those of psychogenic origin, if the psycho-analytic method had not succeeded in effecting an explanation and therapeutic improvement of several cases of the kind* [italics added]. It is not merely permissible, therefore, but incumbent upon us to extend the analytic explanation of melancholia to mania.

What made Freud's views on manic-depression so persuasive was his unparalleled ability to evoke the epic internal struggles in patients with this illness, which he summarized in 1933 in *New Introductory Lectures on Psychoanalysis:*

In this disease . . . the most remarkable characteristic is the way in which the super-ego—you may call it, but in a whisper, the conscience—treats the ego. The melancholiac during periods of health can, like any one else, be more or less severe towards himself; but when he has a melancholic attack, his super-ego becomes over-severe, abuses, humiliates, and ill-treats his unfortunate ego, threatens it with the severest punishments, reproaches it for long forgotten actions which were at the time regarded quite lightly, and behaves as though it had spent the whole interval in amassing complaints and was only waiting for its present increase in strength to bring them forward, and to condemn the ego on their account. . . . It is a very remarkable experience to observe morality, which was ostensibly conferred on us by God and planted deep in our hearts, functioning as a periodical phenomenon. For after a certain number of months the whole moral fuss is at an end, the critical voice of the super-ego is silent, the ego is reinstated, and enjoys once more all the rights of man until the next attack. Indeed in many forms of the malady something exactly the reverse takes place during the intervals; the ego finds itself in an ecstatic state of exaltation, it triumphs, as though the super-ego had lost all its power or had become merged with the ego, and this liberated, maniac [manic] ego gives itself up in a really uninhibited fashion, to the satisfaction of all its desires.

By offering an all-purpose treatment and an all-purpose way of thinking about mental disorders, Freud made Kraepelin's approach to finding specific causes seem hopelessly academic. Though well aware of the distinction between depression and manic-depression (and more inclined than Kraepelin to separate them) Freud's belief that all psychiatric disorders could be understood and treated from a psychoanalytic perspective caused him to lose interest in a system of classification.

SUCH A RETREAT from classification was, in fact, actively promoted by many psychoanalysts at mid-century. Among them was Karl Menninger, one of the most influential American psychiatrists of the period. Along with his father, Charles, and his brother, Will, Karl Menninger had developed the Menninger Clinic in Topeka, Kansas, into a world-famous treatment center in which (as at McLean Hospital) the psychoanalytic perspective was applied to all forms of psychiatric disorders, including schizophrenia and manic-depressive illness. In 1963 he published *The Vital Balance,* in which he explicitly challenged the value of any sort of classification of mental illness, stating:

> Many classifications were composed which endeavor to explain different mental illnesses. . . . From a study of these it seems . . . very evident that there has been a steady trend toward simplification. . . . We have carried this forward to the logical implication that *perhaps there is only one class of mental illness—namely, mental illness* [italics added]. We propose that all the names so solemnly applied to various classical forms and stages and aspects of mental illness in various individuals be discarded.

Menninger was not, of course, denying the existence of distinct patterns of disruptive and dangerous behavior such as mania or depression. Nor was he denying the need to treat them. But if the differences between behavioral disorders were only a matter of degree, and if the treatment—psychotherapy based on psychoanalysis—was ultimately the same, he saw no practical reason to make diagnostic distinctions. And lest you conclude that this view was idiosyncratic, Franz Alexander and Sheldon Selesnick, two psychoanalysts whose influential *History of Psychiatry* was published in 1966, also dismissed Kraepelin as "a rigid and sterile codifier of disease categories; even if these were valid, they contribute to neither understanding of diseases nor their prognosis."

It was left to Thomas Szasz, also a psychiatrist and psychoanalyst, to take this position even further, actually questioning the very idea of mental illness. In *The Myth of Mental Illness* he wrote:

> It is customary to define psychiatry as a medical specialty concerned with the study, diagnosis, and treatment of mental illnesses. This is a worthless and misleading definition. Mental illness is a myth. Psychiatrists are not concerned with mental illnesses and their treatments. In actual practice they deal with personal, social and ethical problems in living.

For a time then, the Kraepelinian approach was in danger of extinction.

BUT THE TIDE would turn. Although the views of Menninger and Szasz would remain in vogue for years, by the early 1970s a renaissance of interest in psychiatric classification was in the making. The leaders were a group of researchers who, like Kraepelin, were convinced that clear diagnostic categorization was essential to understanding the underlying causes of psychiatric disorders. Without a proper system of classification, the hunt for the genetic variations that made some people vulnerable to a particular mental illness would be doomed to failure.

There was also another reason that classification had become so important: drug treatments had been introduced that helped some specific groups of patients but might actually harm others. By the early 1960s, hundreds of thousands of people (including Max) were being treated with imipramine, a drug that, while alleviating depression, may aggravate another mental illness, schizophrenia. Furthermore, similar numbers of people with schizophrenia were being treated with chlorpromazine, a drug that may aggravate major depression. Ignoring diagnostic distinctions, therefore, could lead to the prescription of the wrong drug—with disastrous consequences.

Misapplication of these drugs was particularly likely because they were not developed on the basis of knowledge about the fundamental abnormalities that give rise to mania, depression or schizophrenia. Instead, they were discovered through a series of accidents, of which none was more dramatic than that made in 1948 by a lone psychiatrist, John Cade, working in a provincial hospital in Australia. Like the scientists who are presently hunting for mood genes, Cade was interested in the biological basis of manic-depressive illness. To him it seemed likely that its symptoms reflected the abnormal secretion of a hormone— periods of excessive secretion leading to mania and periods of inadequate secretion leading to depression. But having no clue as to the nature of the hypothetical hormone that might be episodically secreted, Cade decided to look for evidence of this abnormality in the urine of patients who were having a manic attack. Examining urine was, after all, a standard tactic of medical research. It had led Archibald Garrod to the secrets of AKU.

To hunt for a sign in the urine, Cade had a wild idea: he would inject samples of urine from manic patients into guinea pigs and compare their effects with those of samples from normal people. Finding evidence that comparatively small amounts of urine from the manic patients caused seizures in the guinea pigs, Cade became interested in the possible role of a constituent of urine called urate. Because the common salt of urate, sodium urate, is very hard to dissolve, Cade turned to the much more soluble salt, lithium urate (whose great solubility had, as it happens, been reported in 1859 by Archibald Garrod's father, Alfred, in his famous studies of the uric acid deposits in the joints of people with gout). This work led to the unexpected observation that injections of lithium urate made guinea pigs lethargic; and, after testing another lithium salt, lithium carbonate, Cade concluded that it was actually the lithium rather than the urate that was exerting this behavioral effect.

Intrigued by the lethargic guinea pigs, Cade had another wild idea: he would abandon his studies of the urine of his manic patients, and try instead to treat their symptoms with lithium salts. To make sure this was safe, he took some himself. Then, having suffered no ill effects (and knowing that a lithium salt had been widely used by people with heart disease as a substitute for table salt) he gave repeated doses to several patients. Amazingly, it worked.

The first patient to receive this experimental treatment was a fifty-one-year-old man who had been in the hospital for five years with what Cade described as "a state of chronic manic excitement . . . amiably restless, dirty, destructive, mischievous and interfering." But despite the chronicity of his mood disorder, three weeks of treatment with lithium wiped out the mania. With continued treatment the patient became well enough to be discharged from the hospital on a dose of 300 milligrams of lithium carbonate twice daily. Thereafter he returned to his old job; regressed when he stopped taking his pills; and recovered when the lithium carbonate was resumed.

Thrilled by this success, Cade treated more manic-depressive patients with lithium, and had more good results. So too did others around the world who, in systematic clinical trials in the 1950s and 1960s, went on to confirm the considerable efficacy and relative safety of lithium carbonate in treating and preventing recurrences of mania. And even though lithium wasn't always effective against mania, was much less effective against depression, and sometimes had distressing side effects, it revolutionized the treatment of manic-depressive illness—although we still don't really know how it works.

That other mental disorders would also respond favorably to specific pharmaceuticals was demonstrated in the 1950s, with the introduction of a whole series of useful drugs. Among them was chlorpromazine for schizophrenia and both monoamine oxidase inhibitors and tricyclic compounds (which included Max's drug, imipramine) for major depression. With the discoveries of these drugs came an increased interest in the classification of mental illness, as a guide to their prescription. Menninger's idea that "there is only one class of mental illness" would no longer do.

This is not to say that the new drugs were accepted immediately. Many, like Szasz, attacked them as instruments of oppression, chemical straitjackets, a view that Szasz continues to hold. It was, however, a view that would cost Szasz dearly. In 1994, Professional Risk Management Services, the malpractice insurance company that represented Szasz, agreed to pay $650,000 to the widow of one of his patients who was known to have manic-depressive illness. The patient, a fellow psychiatrist, had stopped taking lithium, allegedly on Szasz's advice, and had committed suicide by hanging himself with battery cables.

Having to settle a malpractice suit is only one indication that Szasz's positions have lost favor. Another comes from Alan Stone, my teacher at McLean, who encouraged his students to pay attention to Freud while feeding us Sacher-tortes. Writing the cover story in the January 1997 issue of *Harvard Magazine* (an article called "Where Will Psychoanalysis Survive?" and subtitled "What Remains of Freudianism When Its Scientific Center Crumbles?"), Stone not only comes out in support of using drugs to treat psychological symptoms but also reports his disappointed conclusion that Freud did not really found a scientific psychiatry after all. In his view, "psychoanalysis, both as a theory and as a practice, is an art form that belongs to the humanities and not to the natural sciences."

THE REORIENTATION OF American psychiatry away from psychoanalysis and toward the rest of medicine began picking up steam in the mid-1970s, not long before I met Michael. Reacting to the prevailing opinions of the time, such as those expressed by Szasz and Menninger, groups of psychiatrists at Washington University in St. Louis and Columbia University decided to work together to rekindle interest in the classification of psychiatric disorders as a step in the hunt for their causes.

Their aim followed directly from Kraepelin and Virchow: sharpen the clinical descriptions, so that causes and better treatments can ultimately be found. Because there were no blood tests or urine tests to aid in diagnosis, other types of instruments would have to be developed, especially for purposes of research. To this end standardized testing procedures such as the Schedule for Affective Disorders and Schizophrenia (SADS) were created for structured interviews in which a carefully worded set of questions would be asked. Using such standardized procedures in place of the informal interviews that are used in clinical practice would increase the reliability of the diagnoses because it assures that nothing important is skipped and that all diagnosticians are evaluating each case on the basis of a complete set of facts.

In addition to devising diagnostic instruments for research, these followers of Kraepelin became interested in reestablishing the importance of systematic diagnosis in the clinical practice of psychiatry. To this end they campaigned actively for a revision of the American Psychiatric Association's *Diagnostic and Statistical Manual* (*DSM*), the official handbook of psychiatric diagnoses in the United States. First published in 1952, then revised in 1968, this manual initially defined many disorders in terms of the prevailing psychoanalytic views. With the Kraepelinian renaissance of the 1970s the stage was set for an official shift to a diagnostic scheme that was based on lists of explicit criteria rather than speculations about underlying psychological processes. This came into effect with *DSM-III's* publication in 1980, and was retained in the next complete revision, *DSM-IV*, published in 1994.

But even though many of the framers of *DSM-III* thought of themselves as neo-Kraepelinians, their classification of mood disorders was quite different from the one Kraepelin himself had proposed. He had grouped mood swings of varying severity together because of his hunch that they were all "manifestations of a single morbid process." The viewpoint of *DSM-III,* however, was different: the main focus was on identifying patterns of behavior that permitted reliable classification based on objective criteria, without any speculation about unidentified causes. From this point of view it seemed more reasonable to split up mood disorders into several distinct entities—not because their causes were known to be fundamentally different but because they could be reliably distinguished on the basis of a checklist of features. The framers of *DSM-III* even went so far as to discard Kraepelin's term, "manic–depressive insanity," because of the inclusiveness it had come to imply.

A main originator of the classification of mood disorders adopted in *DSM-III* and *DSM-IV* was Karl Leonhard, professor of psychiatry and neurology at Humboldt University in Berlin, who in 1959 divided these disorders on the basis of what he called their "polarity." To the illness of people who had swings to both the manic and the depressed poles (like Flora) he gave the name *bipolar*. In contrast, the illness of the much larger number of patients who (like Max) had only severe depressions he called *unipolar*, since they swung only to one pole. No matter that these two distinguishable entities sometimes ran in the same families, implying a common cause; the aim was first to establish a basis for reliable clinical distinctions. Hunting for causes would come later.

The framers of the *DSM*s liked Leonhard's distinction between "unipolar" and "bipolar," terms that continue to be useful. Nevertheless, wishing to subdivide mood disorders even further, they came up with a more elaborate scheme that is based on combinations of four basic patterns of symptoms:

A *manic episode* (like those described for Flora and Jerry) is a distinct period of abnormally and persistently elevated, expansive, or irritable mood lasting for at least a week and of such severity that it causes impairment in social or occupational functioning. The mood disturbance must be accompanied by at least three additional symptoms from a list that includes grandiosity, decreased need for sleep, hypersexuality, buying sprees, and impairment of work or relationships. The elevated mood may be infectious, greatly lifting the moods of others, but it is also recognized as excessive by those who know the person well. It may include psychotic features such as delusions or hallucinations.

A *hypomanic episode* is a distinct period of abnormally and persistently elevated, expansive, or irritable mood that lasts at least four days and that is less severe than a manic episode. It does not cause marked impairment of functioning, does not require hospitalization, and does not include psychotic features. In fact, for some individuals, it may be associated with greatly increased efficiency and creativity. Although many of the characteristics of a hypomanic episode may be desirable, it is nonetheless considered evidence of illness because people who have hypomanic episodes are also likely to have destructive manic and depressive symptoms.

A *major depressive episode* (like those described for Max and Flora) includes depressed mood or loss of interest and pleasure in life for at least two weeks along with at least four other symptoms from a list that includes difficulty in thinking, concentrating, or making decisions, sleep disturbances, decreased energy, changes in appetite and weight, feelings of worthlessness or guilt, and recurrent thoughts of death or suicide. Although the degree of impairment varies, to justify such a diagnosis there must be significant distress or some interference in social or occupational functioning.

A *mixed episode* (glimmerings of which are seen in Jerry's story) combines the features of a major depressive episode and a manic episode with rapidly alternating moods of sadness, irritability, and euphoria.

In the course of our lives all of us have experienced many of the feelings that are components of these episodes. We constantly make use of feelings of elation or sadness as signs of how our lives are going and as motivators of change; and we use them to signal this information to others. What distinguishes these normal periods of happiness or sadness from those considered to be components of a "disorder" is their appropriateness to the situation we are in. Normal mood variations are understandable responses to life events; abnormal ones appear to be unprovoked, too intense, too sustained, and may culminate in frantic agitation or paralyzing despair.

Specific patterns of mood disorders are conceptualized in *DSM-IV* as combinations of the four basic types of episodes:

Bipolar disorder, type I: one or more manic or mixed episodes, usually, but not necessarily, accompanied by major depressive episodes (thus Jerry could be classified as bipolar I on the basis of a single manic episode, even though he had never been seriously depressed).

Bipolar disorder, type II: one or more major depressive episodes accompanied by at least one hypomanic episode. (Could this be the correct diagnosis for Flora's father, David?)

Major depressive disorder: one or more major depressive episodes (Max's diagnosis).

Cyclothymic disorder: at least two years of numerous periods of hypomanic symptoms that do not meet the criteria for a manic episode, and numerous periods of depressive symptoms that do not meet the criteria for a major depressive episode (an alternative possibility for David).

Dysthymic disorder: at least two years of depressed mood for more days than not, accompanied by additional depressive symptoms that do not meet criteria for a major depressive episode.

To complicate matters further, *DSM-IV* includes yet another category, *schizoaffective disorder,* to describe the illness of patients who meet the criteria for both schizophrenia and mood disorders. A distinguishing feature is that patients with schizoaffective disorder may have delusions or hallucinations even during periods when their manic or depressive symptoms have subsided, whereas patients with other mood disorders have delusions or hallucinations only in the course of extreme mood swings.

Essentially, then, in *DSM-IV* manic–depressive illness has been split for diagnostic purposes into six subcategories based on criteria that permit their reliable distinction, but this division does not directly address Kraepelin's idea that they may all reflect the same "hereditary taint." So if our goal is to find out if there really is a genetic basis for mood disorders, should we be studying these six subcategories separately? Or might it not be better to consider some or all of these subcategories as alternative manifestations of the same illness?

A WAY TO decide is by determining whether specific subcategories are concentrated in particular families. For example, if bipolar disorder and unipolar disorder ran in different families, that finding would support those who believed it was best to consider them separately. On the other hand, if affected families had a mixture of these mood disorders, that would favor those who, like Kraepelin, believed it was best to consider them together.

To evaluate affected families for this purpose a single member, the "index case," is first identified as having bipolar or unipolar disorder, and first-degree relatives—parents, siblings, and children—are then tested to see if they too have a mood disorder. The reason for sticking with the parents, siblings, and children of a person is that they all share, on average, half of that person's genetic material (though never exactly the same half), whereas all other relatives share less. The results of this analysis were clear: more than a dozen independent studies of first-degree relatives of hundreds of index cases with unipolar or bipolar disorder, summarized in 1986 by Peter McGuffin and Randy Katz, show that these two forms of mood disorder generally run in different families. So it is reasonable to consider these illnesses, at least in some cases, as separate disorders, with different causes.

The main reason to consider bipolar disorder as a separate condition is that first-degree relatives of a bipolar patient have a much higher lifetime risk (8 percent) of being bipolar than do the comparable relatives of a unipolar patient (0.6 percent). In fact, the risk that a unipolar's close relatives will be bipolar is not significantly different from the risk in the population at large (0.5–0.9 percent of both men and women in the United States and Western Europe), whereas the risk to comparable relatives of a bipolar index case is more than ten times as great.

Lest you conclude that there is a complete separation between manic-depression on the one hand and depression on the other, however, the family studies also show an important point of overlap: first-degree relatives of a bipolar patient (such as Flora's brother, Max) also have a greater vulnerability to depression alone. They have twice the lifetime risk of having a major depression than other people have—roughly 10 percent for the relatives of the bipolar patient compared with roughly 5 percent for the population as a whole (about 7 percent in women, 3 percent in men). And lest you think that such a doubling may not be very significant, this same doubling of the lifetime risk of depression is similar to that for first-degree relatives of a unipolar index case, whose lifetime risk is also about 10 percent.

There is yet another wrinkle. A relatively unusual subtype of unipolar disorder, identified by Myrna Weissman and colleagues, called early-onset major depression, strikes in adolescence rather than in later life, when other types of depression become more common. Starting with index cases whose depressive episodes began during adolescence, the risk to first-degree relatives of developing depression at some point in their lives is staggeringly high—about five times the risk to the general population—as shown in the graph on the facing page.

The graph is plotted to show the cumulative risk of having a bout of major depression by a given age. Because it is cumulative, the risk keeps rising for each subgroup, reaching about 5 percent in the general population (with no family history) by the age of sixty. One important result of this study is the finding that the risk to close relatives of an index case who first became depressed after the age of forty is *not much different* from the risk to the population as a whole, indicating that, in general, late-onset depression is not familial (though in some cases it may be). In marked contrast, the risk to close relatives of an index case who first became depressed before the age of twenty is almost 30 percent—making early-onset depression a strikingly familial disorder. But the boundary with bipolar disorder is maintained:

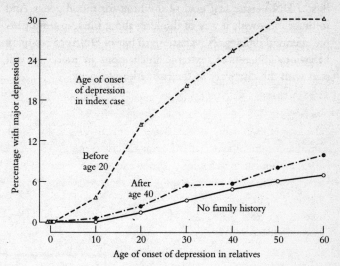

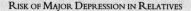

there is about the same amount of bipolar disorder in families with early-onset depression as in the population as a whole.

When taken together, the results of the family and population studies point to sufficient differences in the distribution of unipolar and bipolar disorders, and of a subtype of unipolar disorder with early onset, to justify their separation as distinct entities for the purpose of studies designed to hunt for their genetic basis. And because the lifetime risk to close relatives of patients with bipolar disorder is more than tenfold the lifetime risk to the population as a whole, making it most likely that genes play a critical role in the susceptibility to this distinctive condition, bipolar disorder—especially the most severe form, bipolar disorder, type I—has become the favorite target in the hunt for a genetic basis for mood disorders.

IS THE INCREASED risk of a particular mood disorder in certain families really a reflection of a "hereditary taint"? To prove this with certainty it is necessary to demonstrate the variations in mood genes that are the basis of this inherited vulnerability by actually finding and methodically examining these specific bits of DNA—the first goal of the hunt for mood genes. And in order to provide a way of thinking about the leap from simple chemical differences (variations in bits of DNA) to complex behavioral differences (extreme fluctuations in mood), I will start with the discovery of the basic idea of a gene.

3

THE ASTONISHING LEAP FROM TRAITS TO GENES

The distinguishing traits of two plants can, after all, be caused only by differences in the composition and grouping of the elements existing in dynamic interaction in their primordial cells.

                                        —Gregor Mendel (1865)

When Emil Kraepelin expressed the view that manic-depressive insanity was due to "hereditary taint," he had no way of foreseeing that the specific factors—the genes—that transmit this vulnerability might some day be identified. To Kraepelin, heredity was simply another way of saying that the disorder ran in families, and that he suspected that this trend was caused by nature rather than nurture. Yet in 1865, long before Kraepelin himself began laying the clinical groundwork for a study of the role of genes in mood disorders, the fundamental rules of heredity that would eventually guide this work had already been put forward.

They had been discovered over the course of just a few years by an Augustinian monk, Gregor Mendel, working in his monastery garden in Brno (now in the Czech Republic), and later shown to apply as much to human beings as to plants. The most important of these Mendelian rules is that some inherited properties—such as the colors of the flowers in Mendel's garden or certain human diseases—can be thought of as being controlled by a single unit, which Mendel referred to as an "element of the germinal cell," and to which in 1909 the Danish biologist

Wilhelm Johannsen gave the name "gene." And even though the color of the flower also depends on a host of other factors—including the many genes that control the construction of the flowers to be colored, as well as access to sun, water, and food—Mendel's experiments demonstrated that a single gene can still be fruitfully thought of as controlling the color itself.

BORN IN 1822 to a peasant family in the village of Heinzendorf, then part of Austria, Mendel received the scientific training that was indispensable for his later work only through an ironic misfortune. Admitted to the monastery in Brno as a novice in 1843 and ordained as a priest four years later, Mendel planned to teach natural science at a local school, a common assignment for young priests. But in 1850 he failed the examination for a teaching certificate. To remedy his deficiencies, and to provide him with a more adequate background for teaching—which he would eventually practice—Mendel's abbot sent him to the University of Vienna (later Freud's alma mater). There, from 1851 to 1853, he studied physical and biological sciences with a number of distinguished researchers including Christian Doppler (of the Doppler effect). Legend has it that his interest in plant breeding was stimulated by this same abbot, who believed that improved breeding techniques would aid local farmers. What is known for certain is that in the summer of 1854 Mendel began growing thirty-four strains of peas (obtained from local farm suppliers) to make sure they were inbred lines that produced only uniform plants with specific features of interest. From this survey he selected the strains that he would work with for the next decade.

Domesticated pea plants were a very favorable experimental subject because farmers around Brno had been finding new varieties for centuries; so seeds were available from plants with a number of alternative and easily distinguishable characteristics. Among them were variations in flower color (either purple or white flowers), seed shape (either smooth round seeds or wrinkled shriveled seeds), seed color (either yellow or green), and four other properties (pod color, pod shape, plant height, and plant branching pattern) that Mendel made use of. Starting with strains that he had already tested for constancy (for example, strains in which all the plants have purple flowers and others in which all the plants have white flowers), Mendel began his series of experiments by interbreeding these pure strains to study the properties of the hybrids (for example, he fertilized white flowers with pollen from purple flowers and vice versa). After harvesting and growing the seeds, he observed the color of the flowers in the first generation of offspring.

One possible outcome that comes to mind is that interbreeding plants with white flowers and plants with purple flowers would give offspring with flowers of intermediate color, a light shade of purple caused by dilution of the more intense parental purple with the parental white. But this is not what Mendel found. Instead, he found that interbreeding between pure strains produced flowers that were all the same shade of purple as the purple parent strain, not in the least diluted by the white. With the other six characteristics he studied he found a similar result: one of the alternative characteristics completely dominated the other. In Mendel's own words,

> those traits that pass into hybrid association entirely or
> almost entirely unchanged, thus themselves representing the
> traits of the hybrid, are termed *dominating,* and those that
> become latent in the association, *recessive.* The word
> "recessive" was chosen because the traits so designated
> recede or disappear entirely in the hybrids. . . .

Having obtained hybrids that all had purple flowers, Mendel wondered what had happened to the factor responsible for making white flowers. One possibility was that it was gone forever, somehow permanently obliterated by the factor responsible for making purple flowers. In Mendel's time, this idea seemed perfectly reasonable. It is precisely because Mendel's experiments refuted this idea and replaced it with a profoundly informative alternative that we are still talking about them more than a century later.

To find out what had happened to the factor responsible for the recessive trait, Mendel bred the first-generation offspring plants among themselves. Had the initial interbreeding permanently obliterated the factor responsible for white flowers, there would be no white-flowered plants among the offspring of that mating, the second generation. Instead, Mendel found that many of the second-generation plants had white flowers, indicating that the factor responsible for making white flowers had *not* been permanently lost. In considering the results of this experiment, Mendel realized that he had got much more than he had bargained for: he now had an extremely important clue to the underlying process of heredity.

That clue came from a careful count of the number of second generation plants with purple flowers (705) and the number with white flowers (224). When Mendel pondered these results he was struck by this proportion: about three-quarters were purple-flowered plants and about one-quarter white-flowered plants, a ratio of approximately 3:1. This was clearly no accident: the results of his experiments with the six other traits he studied were all very similar. In each case one of the two forms of each trait completely dominated the other in the first generation (for example, smooth dominated wrinkled; tall dominated short); but in the second generation approximately a quarter of the offspring had the recessive form.

From these results Mendel made a number of deductions, here restated in contemporary language:

1. Certain traits, such as flower color, are controlled by a single, discrete, hereditary determinant: a gene. So even though a pea plant has thousands of different genes that all work together to give rise to this complicated organism, some particular properties can each be thought of as being under the primary control of a single gene.

2. Genes may exist in alternative forms, called alleles (pronounced "al" as in Al Capone, and "eels" as in steals), that we now know to differ in the details of their DNA structure, the basis of hereditary biological diversity. For example, there are two alleles of the gene that controls the color of pea flowers: one, *P* (designated with a capital letter because it is dominant, italicized like all symbols for genes and alleles), gives rise to the purple color of the flowers, while the other, *p* (designated with a lower-case letter because it is recessive), gives rise to the white color of the flowers. Other genes may have more than two alleles, making possible an even greater diversity of forms in the traits they govern.

3. Each individual has two copies of each gene (a gene pair), one member of each pair derived from each parent. The alleles in a particular gene pair (for example, *PP,* or *Pp,* or *pp*) are called a genotype. The outward manifestation of a genotype (for example, purple) is its phenotype (from Greek, meaning "the form that is shown"): the phenotype of the genotype *PP* or *Pp* is the purple color of the flower, and the phenotype of the genotype *pp* is the white color.

4. Sex cells—eggs or sperm—each contain only one member of each gene pair. When sex cells form, the alleles of each gene pair (for example, *Pp*) separate (or "segregate") so that half the sex cells carry one member of the gene pair (*P*), and the other half of the sex cells carry the other member of the gene pair (*p*).

This last point, which assumes all the others, is generally called Mendel's first law, or the law of equal segregation. What it means in human terms is that, of the two members of a gene pair that each parent has to give us, we have an equal chance of receiving either one.

Mendel's amazing findings can be reviewed in such brevity because they have been so well integrated into common parlance. Yet there is one critical word—*allele*—that still causes some eyes to glaze over. Many people prefer to avoid it altogether and instead attempt to expand the definition of the word "genes" by also using it to mean "alternative forms of genes," as in "some flowers are purple and others are white because they have different genes." But such a use is misleading. Plants do indeed have different genes, such as a "flower-color gene" and a "seed-color gene." Each of these genes encodes a completely different protein that controls a different property. In contrast, what accounts for the different colors of Mendel's flowers are *variations* in the "flower-color gene"—that is, alleles—so that one variation of this gene (the allele *P*) encodes a protein that is slightly different from that encoded by another variation (the allele *p*). Inherited human differences are likewise produced by *alternative forms* of genes—alleles.

Put another way, what distinguishes you and me is not different genes, because all human beings have the *same genes*. Instead what distinguishes us genetically is that we have each inherited quite a few *different alleles*. So the difference between the word "gene" and the word "allele" is critical. And rather than misuse "genes," or continue to resort to more cumbersome phrases such as "alternative forms of genes" or "gene variants," I will stick with "alleles," the shortened form of "allelomorphs" (from Greek words meaning "alternative forms") that was coined for this purpose early in the twentieth century by the British geneticist William Bateson.

To help relate Mendel's findings to cases of human inheritance, they can be recast in the family-tree format called a pedigree, like the one that Michael sent me. To this end, consider the offspring of a mating between a white-flowered plant with two identical alleles (*pp*), and a purple-flowered plant with two different alleles (*Pp*). The color of the flowers (the phenotype) is indicated by a filled-in symbol for purple flowers and an open symbol for white flowers. The results of this mating are shown in the diagram on the facing page. The genotype of each individual appears beneath each symbol.

From Mendel's first law we would expect that, on the average, half the offspring would inherit a *P* allele from the purple parent and that all the offspring would inherit a *p* allele from the white parent. As a result, about half the offspring would have the *Pp* genotype and a purple phenotype, and the other half would have the *pp* genotype and a white phenotype. In the example in

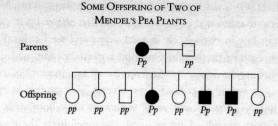

SOME OFFSPRING OF TWO OF
MENDEL'S PEA PLANTS

the diagram above, chance segregation gave three of eight plants with purple flowers, rather than the hypothetical four of eight.

The example also illustrates that (to use a human analogy) if the order of birth were from left to right, then the first, second, and third offspring would have all been white. If the parents had stopped having children after the first three they would have had no purple offspring: half is just the expected average.

This neat and simple picture is not, of course, the whole story of heredity. It is only found in cases in which a single gene controls a specific trait, such as the color of pea flowers. Such traits—which are called single-gene traits, monogenic traits, or Mendelian traits—were the ideal experimental subjects that Mendel relied on to be able to infer the idea of the gene as a fundamental unit. From them has come the often used phrase, "one gene–one trait." But most human traits that we are interested in, such as mood disorders, are thought to be polygenic ("from many genes"): they are believed to be influenced by *more than one* gene because pedigrees depicting their transmission from generation to generation do not show the simple patterns that Mendel found. When we later go on to consider pedigrees of families with members who do have single-gene diseases, it will become apparent how much Michael's pedigree, like that of other families with manic-depressive illness, deviates from such monogenic patterns. Yet, despite their greater complexity, Mendel himself already laid the groundwork for analyzing traits that are influenced by more than one gene—the first of three "genetic more-than-one"s that I will introduce with examples that can in fact be derived from studies with Mendel's peas.

FOLLOWING HIS INITIAL experiment with single traits, Mendel became interested in breeding plants that differed not just in one trait but in two—such as seed shape *and* seed color. Even though each of these traits could be separately examined in all the offspring, he decided to find out how combinations of shape and color would distribute over several generations. Starting with highly inbred plants, some with smooth yellow seeds and others with wrinkled green seeds, Mendel bred them with each other, knowing full well from his earlier experiments that since the traits smooth and yellow are dominant, all the first-generation seeds would be both smooth and yellow. The point of the experiment was to find out what happened in the *next* generation, when he interbred the plants from the first-generation seeds.

One possibility was that the second-generation seeds would each be like one of the grandparents—either smooth and yellow or wrinkled and green. The alternative was that the seed-shape gene and the seed-color gene had no relationship to each other, so that the two alleles of one gene could mix freely with the two alleles of the other. Were this the case some second-generation seeds would show two new mixtures not found in the grandparents: smooth and green seeds, and wrinkled and yellow seeds.

The results supported the second possibility: free mixing to form new combinations. Once again there were well-defined proportions. Of every sixteen offspring, on the average, nine were smooth and yellow; three were smooth and green; three were wrinkled and yellow; and one was wrinkled and green. All these combinations fit with the established fact that smooth and yellow are each a dominant trait. From these results Mendel concluded that the seed-shape gene behaves independently of the seed-color gene. This and similar results with the other traits that he studied form the basis for Mendel's second law, the law of independent assortment: during sex-cell formation the segregation—that is, the distribution—of one gene's alleles to a particular sex cell is independent of the segregation of another gene's alleles into that sex cell.

The upshot of Mendel's second law is that a large number of distinctive pea plants can be generated with different combinations of characteristics (tall with white flowers and smooth yellow seeds; short with white flowers and wrinkled green seeds, and so on). As Mendel himself pointed out, it is possible to have 128 different progeny from mixtures of the two alternatives of the seven traits that he examined ($2 \times 2 \times 2 \times 2 \times 2 \times 2 \times 2 = 128$). Such mixtures form the basis for particular varieties of pea plants that are of interest to farmers and gardeners.

Recognizing the importance of such mixtures can serve as a bridge between the simple monogenic traits that Mendel began with and the polygenic traits that are our major concern. When thinking about peas as experimental subjects it may seem obvious to pick out the color of the peas and the texture of the peas as separate traits. But from an alternative perspective these two attributes of a pea could easily be merged to form a single entity. One way to appreciate this is to stop thinking about pea seeds as experimental subjects and to start thinking of them as edible commodities. When viewed in this way the trait of interest might be neither a color nor a texture alone, but instead an overarching property such as "desirable-edible-seed-variety." And the overarching property might combine both the features green and wrinkled, and thus be controlled not by one gene but by two. If the fashion were to prefer such peas, they might take on a special name such as "grinkled" to define a property controlled by the combined actions of alleles of two genes.

In this example we happen to be able to break down the trait of interest ("desirable-edible-seed-variety") into two equally important subtraits (color and texture), each of which is controlled by a single gene. But most polygenic traits—human height is one—are influenced by alleles of multiple genes, and their effects cannot usually be dissected into observable components. Because of the multiplicity of the possible interactions between alleles of many genes and the environmental factors that together influence traits such as human height, these traits are often referred to as complex traits.

Though some unusual alleles may exert such profound effects on a complex trait that they *alone* may be the major factor in producing an extreme phenotype (as in the case of human giants or dwarfs), the specific effects of most alleles that influence complex traits are very difficult to pick out in this way.

In fact, for many years after Mendel, geneticists threw up their hands at the prospect of identifying most of the genes involved in complex traits, because the Mendelian approach required that their effects could be inferred on the basis of the phenotypes—the outward manifestations. There was just no way to find and study these genes as isolated chemical entities in order to determine more directly their fundamental biological effects. But, as we will soon see, contemporary DNA technology has changed all this. Right now every detail of each human gene is being systematically scrutinized, base pair by base pair. The outcome is an accumulating knowledge about human genes that makes it possible to hunt them down and figure out their contributions to complex traits in ways that Mendel could never have imagined.

BUT TWO OTHER complexities that bear on mood genes—two other "genetic more-than-one"s (as shown in the table on the facing page)—are already understandable through a Mendelian approach, without elaborate studies of human DNA. Such understanding follows easily from further work with the same peas with which Mendel made his revolutionary discoveries. The second "genetic more-than-one" is that *more than one* external biological property (phenotype) may be controlled by alleles of only one gene, a common feature of gene action called pleiotropy (from Greek words meaning "more turnings"). This point becomes clear once we recognize that the real reason farmers decided to cultivate wrinkled peas was not a preference for their shriveled appearance but for their considerably sweeter taste. In this case we even know *how* an allele of one gene can control the very different properties of wrinkling and sweetness, and the story is very instructive.

THREE "GENETIC MORE-THAN-ONE"S

Term	Meaning	Example
Polygenic	More than one gene, *all acting together* to control one phenotype	The *combined* actions of alleles of two genes give rise to "grinkled" (green and wrinkled) peas
Pleiotropic	More than one phenotype controlled by a single gene	Alleles of one gene give rise to *both* sweetness and wrinkling in peas
Genetically heterogeneous	More than one gene, *each* controlling the same phenotype	An allele of *any one of* several genes can give rise to wrinkled peas

Like all genes, the one in question encodes a protein, in this case a protein that helps convert the pea's sugar into starch. It has two alleles: an allele that encodes an active form of this protein and yields smooth peas, and an allele that makes a defective protein and yields wrinkled peas. In peas with the defective protein the sugar is not converted to starch, and a great deal of sugar accumulates, making the peas very sweet. The accumulation of sugar as the peas develop leads them to retain excess water (the sugar molecules hold on to water); their skins expand to keep from bursting, and they become extremely plump and swollen. Then, as the peas mature, they begin to dry out, shrinking away from their expanded skins as they wrinkle. The result is two phenotypes—one affecting the pea's appearance and the other its sweetness—from the same allele of a single gene.

Recognizing that even one of Mendel's seemingly simple genes actually works in unexpected and complicated ways, we should not be surprised if the alleles of genes that on the face of it have no obvious relationship to any brain function may turn out to play critical roles in mood disorders. Just as the wrinkled phenotype is an unexpected, mechanical consequence of a defect in the normal biological function of a gene involved in the conversion of sugar to starch, so too may the phenotype that is mood disorder be an unexpected consequence of a defect in the normal, non-emotional, biological function of mood genes. Put simply, many actions of mood genes may have nothing to do with mood.

The third "genetic more-than-one" that can be derived from extensions of Mendel's work is that alleles of more than one *different* gene can *each* give rise to the same phenotype. This too is the case in wrinkled sweet peas, in which different genes each encode a protein that controls a different step in the sequential conversion of sugar to starch. As a result, alleles that encode a defective protein *at any of these biochemical steps* block the whole assembly line in the manufacture of starch, leading to the accumulation of sugar and to wrinkling.

Like the two other "genetic more-than-one"s, this genetic property, called genetic heterogeneity, is extremely common. For example, many human genetic disorders that produce extreme shortness do so by influencing different biochemical steps that contribute to the complex trait of height. One such disorder, achondroplasia, caused by a defect in one gene, afflicted most of the actors who played Munchkins in the 1939 movie, *The Wizard of Oz*; another, pycnodysostosis, caused by a defect in a different gene, crippled the famous French painter, Toulouse-Lautrec. Known defects in a number of other genes can also each cause extreme shortness. Therefore, it should not be surprising if it is found that alleles of *more than one* different gene working independently (genetic heterogeneity) or sometimes all in concert (polygenic inheritance) influence the vulnerability to manic-depressive illness. As with many other complex traits, all three "genetic more-than-one"s are likely to be operative in manic-depression.

The final lesson we can learn from wrinkled peas concerns the naming of genes. Gene names tend to be based on an unusual phenotype that is associated with a particular allele, rather than on the normal function of the protein that the gene encodes, which is rarely known when the gene's effect is first detected. In the case of wrinkled peas, the controlling gene has been named *rugosus* (Latin for "wrinkled") for a distinctive overall appearance rather than, say, *starchy* for the specific action of the protein it encodes. And the same thing applies to such names as "manic-depressive-illness gene" or "mood-disorder gene" or the more cryptic "mood gene" for a gene with alleles that predispose people to develop the phenotype called manic-depressive illness. For now we can call them mood genes because mood disorders are the phenotype we want to explain. Only when mood genes are actually identified as specific bits of DNA with specific structures will their range of functions, like those of *rugosus*, become more apparent.

The results of Mendel's eight years of research, the foundation for all of genetics, were summarized in a report, *Experiments on Plant Hybrids*, that he presented at meetings of the Brno Society for the Study of Natural Science on February 8 and March 8, 1865, and published in German in the society's *Transactions* a year later. But despite the compelling nature and profound implications of Mendel's revolutionary paper, it did not generate much enthusiasm at the time. Here is how the response to its initial presentation was recreated by Loren Eiseley:

> Forty people were present in the room at the schoolhouse where the lecture was given. They were not ignorant people. Botanists, a chemist, an astronomer, a geologist were among those present. In the next month Mendel spoke again to the same audience recounting before them his new theory upon the nature of inheritance. The audience listened patiently. At the end of the blue-eyed priest's eager presentation of his researches, the still existing minutes of the society indicate that there was no discussion. . . .
>
> No one had ventured a question, not a single heartbeat had quickened. In the little schoolroom one of the greatest scientific discoveries of the nineteenth century had just been enunciated by a professional teacher with an elaborate array of evidence. Not a solitary soul had understood him.

The published version of this lecture met with the same lack of interest. Despite the fact that the *Transactions of the Brno Society for the Study of Natural Science* was distributed to many European scholars and libraries, Mendel's paper had no impact at the time. Charles Darwin is known to have received a copy, but he paid no attention (though Mendel is known to have read and annotated his copy of Darwin's *Origin of Species*, which had been published in 1859, and which Mendel's work would eventually help elucidate and support). Mendel tried to gain the attention of other scientists, especially the esteemed botanist Carl Nägeli, but they too ignored him. There is, in fact, only one reference to Mendel's paper in the scientific literature before 1900, and that was in the context of plant biology rather than genetics. Since Mendel's brief research career ended when he himself became abbot in 1868 (though he did publish his ideas about tornados, also in the *Transactions of the Brno Society*, after one passed over his monastery in 1870), his elegant discoveries seemed destined for obscurity.

BUT IN 1900, sixteen years after his death, Mendel became famous. In that year three European botanists made discoveries about heredity like those that Mendel had already reported; and all three found and acknowledged his thirty-four-year-old publication. Mendel's paper was also unearthed by William Bateson, whose English translation was published by the Royal Horticultural Society of London in 1901, and who (in addition to coining the valuable word "allelomorph") coined the term "genetics" to name the new science of heredity.

What particularly excited Bateson was not only the beauty of Mendel's botanical studies but also their broader implications. Within a few years Bateson went on to show, in studies with chickens, that the same rules Mendel had proposed for plants also applied to animals. And, as we shall now see, Bateson actively promoted their extension to all aspects of human heredity, thus helping to lay the groundwork for the hunt for mood genes. In fact, the first evidence that Mendel's discoveries also held for people came from their applicability to inherited human diseases.

4

FROM PEAS TO PEOPLE

Man is, in many ways, very unsuitable as an object for the study of genetics. Families are too small for dependable determinations of ratios, desired test matings cannot be made, and study of more than a very few generations for any particular purpose is not often possible. The social implications of human genetics are so great, however, that this subject must be investigated; and there are some real advantages in the material. . . .

—Alfred Sturtevant (1965)

In every case of every malady there are two sets of factors at work in the formation of the morbid picture, namely internal or constitutional factors inherent in the sufferer and usually inherited from his forebearers, and external ones that fire the train.

—Archibald Garrod (1931)

The reason William Bateson could so quickly make the leap from peas to people is that he knew about Archibald Garrod's work with alkaptonuria (AKU), the black urine and black joint disease that I first mentioned in discussing Michael's family. In making this leap Bateson helped put the seemingly mysterious processes of human heredity on a scientific footing by showing that certain human traits, such as AKU, were transmitted as reliably as the colors of flowers. For Bateson had realized that the only difference between the study of heredity in peas and in people is that peas could be bred in whatever combinations and numbers the experimenter desired, whereas human matings were arranged in other ways. But the rules of hereditary transmission were the same. Just as single genes controlled certain specific pea traits, such as flower color ("one gene–one trait"), so too could single genes control certain human diseases such as AKU ("one gene–one disease").

What made AKU such a marvelous starting point for this new approach to human genetics is that its most puzzling feature—the particular way that it ran in families—could be immediately solved by the application of Mendel's first law. Garrod had found a number of families in which more than one sibling had AKU (his clue that the disorder was hereditary) but, surprisingly, *no other affected relatives*. Not one of the grandparents or parents of his AKU patients showed any sign of the disease. Nor did any of his patients' children. How could AKU be hereditary in the absence of evidence of transmission from generation to generation?

Without knowing it, Garrod had found an important clue to the answer. He had noticed that the parents of some of his patients were cousins, a finding that fit with the long-standing observation that children of such consanguineous ("common blood") marriages have a much greater chance of developing rare hereditary diseases—presumably one reason for cultural restrictions on marriage among kin. But in the absence of a fundamental understanding of the hereditary mechanism, Garrod didn't realize the significance of this finding. Nor did Bateson.

Not until he read Mendel's paper. Once he did, Bateson recognized that Garrod's observations must mean that AKU is caused by inheritance of a pair of recessive alleles (which I will call *a*), and that the parents of Garrod's AKU patients must each have one *a* allele and one normal allele (which I will call *A*). Because *A* is dominant, the parents show no evidence of abnormality. But (from Mendel's first law) each of their children would have one chance in two of getting an *a* from one of the parents, and one chance in four of getting two *a*s—an *a* from each parent. With two *a*s and no *A*s, a child would have the phenotype called AKU. And so too would (on the average) one in four unlucky siblings. Because it appears only with the inheritance of a pair of recessive alleles (*aa*), AKU is called a recessive disease.

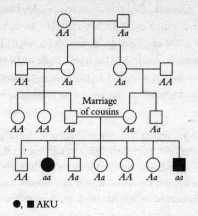

●, ■ AKU

Garrod's finding that his patients with AKU (who are *aa*) did not transmit the disease to their children could also be explained by Mendel's first law. It was only necessary to recognize that the absence of the disease in these offspring was a consequence of the rarity of the *a* allele in the general population. Given this rarity (and assuming random mating), AKU patients would almost always marry people with two normal alleles (*AA*). As a result their children would all be *Aa*, inheriting an *A* from the unaffected parent and an *a* from the affected parent. Being *Aa*, none of these children would show signs of AKU.

Bateson could also explain why the parents of AKU patients were often cousins by proposing that the rare *a* allele had made its way into an AKU family by way of a shared ancestor who, unknowingly, carried and transmitted it. The descendants who inherited a single *a* allele showed no sign of its presence. Only by the marriage of two cousins (genotype *Aa*), who both carried their ancestor's *a*, could the defect finally declare its presence among their children. Such transmission of *a* alleles via parents who are cousins is illustrated in the diagram above.

In December 1901 Bateson presented a report to the Evolution Committee of the Royal Society, explaining these ideas.

In the ensuing year Bateson, the forty-year-old Cambridge biologist, and Garrod, the forty-four-year-old London physician, exchanged a number of letters that led to the publication in December 1902 of Garrod's landmark paper, "The Incidence of Alkaptonuria: A Study in Chemical Individuality," in which he too announced the solution to the long-standing mystery of the role of consanguineous marriages in passing on rare hereditary diseases:

> There is no reason to suppose that mere consanguinity of the parents can originate such a condition as alkaptonuria in their offspring, and we must seek an explanation in some peculiarity of the parents, which may remain latent for generations, but which has the best chance of asserting itself in the offspring of the union of two members of a family in which it is transmitted. This applies equally to other examples of that peculiar form of heredity which has long been a puzzle to investigators of such subjects, which results in the appearance in several collateral members of a family [that is, siblings] of a peculiarity which has not been manifested at least in recent preceding generations. It has recently been pointed out by Bateson that the law of heredity discovered by Mendel offers a reasonable account of such phenomena.

Had Garrod stopped there he would have already made an enormous contribution to our understanding of inherited human diseases. But there was more. Garrod was able not only to relate AKU to a single gene, but also to relate both gene and disease to something quite different: a particular type of protein called an enzyme. Proteins of this kind, which control all the chemical conversions in the body, from the digestion of food to the manufacture of brain molecules that influence our moods, proved to be the priceless lead to an understanding of the relationship between genes and their biological and physiological functions, later solidified by way of the Central Dogma.

IT WOULD BE difficult to overestimate the importance of Garrod's bringing together three such seemingly unrelated entities—genes, enzymes, and phenotypes (in this case, diseases). Mendel had already brought together two, by offering evidence that a single gene controls the color of a flower. So too had Bateson, by pointing out that a single gene controls the development of at least one disease. But how does this effect of a gene actually occur? By what series of steps is a gene translated into an observable property of a living creature? Neither Mendel nor Bateson had any idea.

The clue that led Garrod to relate (albeit vaguely) a gene to an enzyme was the urinary pigment that his patients excreted. In the 1890s chemical studies had shown that this black pigment was an oxidized form of homogentisic acid, a chemical that all of us manufacture. Normally this chemical is broken down to water and carbon dioxide, and the latter is excreted by the lungs. But in people with AKU, homogentisic acid is not degraded in this way. Instead, it accumulates in the body, where some of it is converted to the black pigment that is deposited in cartilage and other tissues, eventually resulting in the distinctive black joints and signs of arthritis that Rudolf Virchow first observed. The rest is excreted in the urine, where it may be oxidized to give the dark color. But how does the degradation of this chemical depend on inheritance of particular alleles of a gene?

Although it would take roughly half a century to answer this question properly, Garrod was in a position to make a good guess: the normal allele must, in some way, work by means of an enzyme that breaks down homogentisic acid. What made this an informed guess rather than a wild idea is that enzymes had already been discovered in the middle of the nineteenth century as the discrete components of brewer's yeast that are involved in beer fermentation and the leavening of bread (the word "enzyme," introduced in 1878, is derived from the Greek "leaven"). In this process each enzyme controls a step in the conversion of sugar into alcohol, chopping or modifying the molecule of sugar until only alcohol, water, and bubbles of carbon dioxide remain. The sum of these stepwise transformations is a metabolic pathway, a pathway made up of a precise sequence of biochemical steps, each controlled by an enzyme. And just as yeast cells employ a series of enzymes in metabolic pathways, sequentially breaking down or building up their internal chemicals for many purposes, so too do we.

From a rudimentary form of this general idea Garrod concluded that homogentisic acid was manufactured as a component of a metabolic pathway and that there must be an enzyme (later identified and named homogentisic acid oxidase) that normally converts it to something else. Furthermore the gene responsible for AKU must in some way control this enzyme, so that the normal allele gives rise to normal enzyme whereas the defective allele gives rise to defective enzyme. And the clinical data indicated that inheritance of only one normal (A) allele provided sufficient amounts of normal enzyme for the metabolic pathway to continue to operate, just as the inheritance of just one purple (P) allele was sufficient to make Mendel's flowers purple. Like Mendel's P allele, the A allele is dominant; and like Mendel's p allele, the a allele is recessive.

Since Garrod's time we have learned that enzymes are, in fact, specific proteins, each with a specific function. We have also learned that alleles of genes are actually alternative sequences of DNA that encode alternative protein structures, so that A gives rise to a protein that acts as an enzyme, whereas a gives rise to an alternative form of this protein that cannot do this job. We now also know that not all the proteins encoded by genes are enzymes: many proteins have other biological functions such as receiving chemical signals (for example, as a brain receptor for dopamine) or transporting certain chemicals (for example, as a brain transporter for serotonin). Although Garrod did not understand these molecular details, he had grasped the fundamental relationship of specific alleles to enzyme (or other protein) function: one allele makes a normal enzyme (or other protein), whereas another may make a defective one. Garrod also came to recognize that the enormous variability of human characteristics, including behavior, was related to differences in what he called our "chemical individuality," the internal chemical differences ultimately traceable to different mixtures of alleles.

In 1909 Garrod published a now classic book, *Inborn Errors of Metabolism,* that elaborated on this view of AKU and extended it to several other diseases. Among them was albinism, which also reflects the lack of a functioning enzyme (in this case, an enzyme needed to make skin pigment) because of the inheritance of two defective recessive alleles of a single gene. Over the years further examples, such as sickle cell anemia and cystic fibrosis, were discovered, all of them transmitted in the Mendelian recessive pattern that Bateson had figured out. In these cases, however, consanguineous marriages are not required to give rise to affected children, since the alleles are quite common in certain populations (for reasons that I will come to later). Because they are caused by alleles of a single gene they are frequently referred to as monogenic or Mendelian diseases—classic examples of the dictum "one gene—one disease"; and the genes and alleles involved are generally called "causative."

But not all diseases that reflect an abnormality in a single causative gene show the pattern of inheritance of AKU. In fact, a number of hereditary diseases that were already known before Garrod's and Bateson's work on AKU show a different pattern, called dominant inheritance. In these cases the abnormal allele is dominant, like Mendel's *P* allele that gave purple flowers, so that transmission from only one parent leads to the disease. Among these Mendelian dominant diseases, one has played an important role in the development of psychiatric genetics, as the first psychiatric disorder in which the causative gene was found on the basis of direct studies of DNA. Discovered in 1872, only seven years after Mendel's lectures, it is called Huntington's disease.

HUNTINGTON'S DISEASE WAS named for George Huntington, the twenty-one-year-old physician who identified it in several families who lived in and around East Hampton, New York. These unfortunate families had already been noticed by Huntington's father, who, like Garrod's, was also a physician. In retrospect it is very likely that all the affected people whom Huntington examined were descendants of a single ancestor who had settled on Long Island many generations earlier. As we shall see, the existence of such ancestors, called founders, in small, relatively isolated communities, may greatly aid the search for disease-causing genes.

Huntington called the disease "hereditary chorea" (the name from the same root as "choreography") because the patients often made dancelike body movements that they could not control. He summarized its salient features in a brief paper:

> There are three marked peculiarities in this disease: 1) its hereditary nature, 2) a tendency to insanity and suicide, and 3) its manifesting itself as a grave disease only in adult life.
>
> 1. Of its hereditary nature. When either or both the parents have shown manifestations of the disease, and more especially when these manifestations have been of a serious nature, one or more of the offspring almost invariably suffer from the disease if they live to adult age. But if by any chance these children go through life without it, the thread is broken and the grandchildren and great-grandchildren of the original sufferers may rest assured that they are free of the disease . . .
>
> 2. The tendency to insanity, and sometimes that form of insanity which leads to suicide, is marked. I know of several instances of suicide in people suffering from this form of chorea, or who belonged to families in which the disease existed . . .
>
> 3. Its third peculiarity is its coming on at least as a grave disease only in adult life. I do not know of a single case that has shown any marked signs of chorea before the age of thirty or forty years, while those that passed the fortieth year without symptoms of the disease are seldom attacked. . . .

The insanity that Huntington described includes serious depression, which can be a major presenting symptom of this tragic disorder, and sometimes mania as well. Like the abnormal movements, the depression can often be alleviated by palliative medications. But the tragic course of the disease cannot yet be halted, despite a growing understanding of its underlying cause. For even though, as I will show later, the unusual alleles that are responsible for Huntington's disease have been isolated, an effective treatment has not yet been developed.

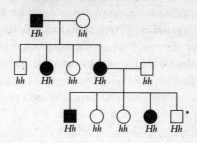

INHERITANCE OF HUNTINGTON'S DISEASE
(DOMINANT INHERITANCE)

●, ■ Huntington's disease

*This individual is too young to show
signs of Huntington's disease

That Huntington's disease is transmitted by a single dominant allele (*H*) is indicated by its pattern of inheritance (shown in the diagram above), which contrasts strikingly with that of AKU. Although inheritance of a single *H* allele invariably leads to the appearance of the disease, the age at which it strikes varies greatly. This explains why one person in the pedigree (marked by an asterisk) who has inherited the *H* allele does not yet display any symptoms. And this defective allele may remain in hiding for many more years—even longer than Huntington believed. In fact, about 5 percent of people who inherit it show no symptoms until their seventies. So we see that even in this classic example of a causative gene there are additional factors that control the appearance of the phenotype—an observation that can help shape our thinking about Michael's family.

THE DESCRIPTION MICHAEL sent me of the intergenerational pattern of mood disorders in his family was in the pedigree format used to describe diseases that are controlled by a single gene. But if we compare the patterns of transmission of monogenic diseases—with AKU as the prototype of a Mendelian recessive disease and Huntington's disease as the prototype of a Mendelian dominant disease—we see that the pattern of manic-depressive illness in Michael's family is not typical of either alternative. It doesn't fit the typical pattern of rare recessive diseases that Bateson and Garrod had deciphered, because there is transmission down the generations from grandmother (Flora) to grandson (Jerry). Nor does it fit the typical pattern of dominant diseases, because the only way that Jerry could have inherited it from Flora was by having it pass through Michael—yet Michael himself showed no signs of a serious mood disorder. So what can we conclude from Michael's pedigree? What could account for the atypical pattern in this family and others with manic-depressive illness, in which there is frequently transmission down the generations, as with dominant diseases, but also sometimes affected children but no affected parents, as with recessive diseases? Does this atypical pattern demolish our working idea that genes play a significant role in manic-depressive illness?

Not at all. Many common familial diseases, such as diabetes and high blood pressure, show patterns of transmission that don't fit with those Mendel described. These diseases, called complex diseases, reflect the combined effects of two types of factors: alleles of several "susceptibility" genes, none of which is sufficiently influential to be called causative; and aspects of environment such as diet and cultural practices, as well as unknown circumstances. There are also diseases that are in between the Mendelian and the complex: these disorders are dependent on an abnormality in a *single* gene but with a pattern of appearance among a group of relatives that is as puzzling as that found in Michael's family.

A notable example of an in-between case is a disease called acute intermittent porphyria, which can be transmitted from one parent to a child via a single allele, in this case (I)—the defining feature of a dominant disease. But in striking contrast to Huntington's disease, in which everyone who inherits the abnormal allele will eventually show signs of the disease, in the case of porphyria only about one person in ten who inherits it becomes sick. The others, nine people out of ten, have no symptoms—none! Yet we know from direct genetic and chemical testing that they too have inherited the abnormal allele. And all of them can transmit the full-blown illness to their children.

Acute intermittent porphyria (which I will henceforth call simply porphyria) has recently received a good deal of attention because it is believed to have been a cause of the idiosyncratic behavior of King George III of England, the subject of the movie *The Madness of King George*. Like George, many patients with porphyria have mental symptoms—usually anxiety or depression, but sometimes hallucinations, delusions, and even mania, of such great severity that they may lead to long-term hospitalization. Therefore it is not surprising that until Ida Macalpine and Richard Hunter assembled evidence that the source of George's bizarre conduct was probably porphyria, a popular retrospective explanation for George's madness was manic-depressive illness. In their book *George III and the Mad Business,* Macalpine and Hunter cite as an example of this explanation the 1941 view of the American psychiatrist Manfred Guttmacher:

> The mental disorder which seized George III on five
> separate occasions was manic-depressive insanity. . . . From
> my analysis of the events preceding George's attacks of
> insanity, and of the illnesses themselves, it appears that
> frustration was the major force behind his disorders. . . .
> Believing that, as a king, he should be all-powerful, he
> became unbalanced when he found himself impotent and
> unable to act. . . . Self-blame, indecision and frustration . . .
> destroyed the sanity of George III. Had it been his lot to
> be a country squire, he would, in all probability, not have
> been psychotic.

Unpersuaded by Guttmacher's speculations, Macalpine and Hunter opted instead for porphyria because of several features of George's illness. For one thing, his episodes of madness were accompanied by acute abdominal pain and muscle weakness, both of which are typical of porphyria attacks. These symptoms are found together because they reflect the same immediate cause—the accumulation in the body of the toxic chemical porphobilinogen, which, by transiently damaging nerve cells and brain cells, gives rise to the disorder's physical and mental abnormalities. Also typical of porphyria is that (as featured in the movie) George's attacks were accompanied by the excretion in his urine of a port-wine–colored substance, presumably porphobilinogen.

As with AKU, the identification of the chemical responsible for the colored urine in porphyria provided an invaluable clue to the function of the defective gene. We now know that porphobilinogen is manufactured in the body as a step in the formation of a chemical called heme, which gives the hemoglobin in blood its red color and is directly involved in the transport of oxygen. We also know that a critical step in the manufacture of heme is controlled by an enzyme called "porphobilinogen deaminase," and that the gene which encodes it has a normal allele (*i*) which produces active enzyme and an abnormal one (*I*) which produces inactive enzyme. Having just one normal allele, which produces half the normal level of enzyme, however, is quite enough to make the necessary amounts of heme and hemoglobin.

Problems arise only if there is such a marked increase in the production of the porphobilinogen that the half-normal levels of the enzyme can't handle the load. This results in a buildup of the chemical to the toxic levels that injure nerve cells, producing an acute attack. As this buildup continues, the bulk of the extra chemical spills into the urine, imparting the port-wine color.

What causes the overload? Known factors include psychological stress and drugs, such as barbiturates, which both stimulate the manufacture of porphobilinogen. But there are also unknown environmental factors (the "environmental background") as well as unidentified alleles of other genes (the "genetic background") that influence the individual's response to stress or drugs. It is variations in both the environmental and genetic backgrounds that presumably determine whether or not a person who has inherited a defective allele (*I*) will develop symptoms of porphyria.

The point of describing porphyria is to provide a memorable example of a property of genes: that in some individuals certain alleles (such as *I*) may lurk beneath the surface and *never* burst out into the open as, in this case, an attack of madness, pain, and port-wine urine. Geneticists have a word—"penetrance"—to describe whether or not an allele penetrates to the surface to declare itself. In those people with *I* who have attacks of porphyria this allele is said to be "penetrant." In those with *I* but no attacks the allele is said to be "non-penetrant."

A simple way of describing penetrance is as the percentage of people with the defective allele who have symptoms of the corresponding disorder. In the case of porphyria, in which only one out of ten people who inherit *I* show any signs of the disease, the defective allele is said to be 10 percent penetrant. This proportion fits the fact that of George III's family, only one of his sixteen children (Augustus, Duke of Sussex) is known to have had the typical symptoms accompanied by dark red urine. In other words, of the eight children out of sixteen who would have been expected to have inherited *I,* in only one of those children was *I* penetrant. In contrast, in the case of Huntington's disease, in which everyone who inherits the abnormal allele eventually develops symptoms, the allele is said to be 100 percent penetrant.

THE EXAMPLE OF PORPHYRIA suggests a solution to the puzzle raised by Michael's pedigree: how could Michael have transmitted a dangerous mood gene allele (or alleles) from his mother, Flora, to his son, Jerry, without himself showing clear signs of manic-depressive illness? Based on the precedent of porphyria, the answer might be that Michael does indeed carry the critical allele (or alleles) that increases susceptibility to manic-depression but is unaffected because he lacks the modulating effects of genetic and environmental background that are needed for it to be penetrant. It even raises the possibility that the mood disorder in Michael's family is transmitted by a single dominant allele that is penetrant in Flora and Jerry (and possibly Max) and non-penetrant in Michael. Some researchers studying the distribution of manic-depression among relatives have in fact concluded that a single dominant allele of a mood gene plays a decisive role in the transmission of this disease—

though others argue that alleles of at least three mood genes must work together to give rise to this vulnerability.

But what about David, Flora's father, whose pattern of behavior Michael had found so difficult to categorize? Because David's symptoms were hardly serious enough to warrant a diagnosis of manic-depressive illness we would have to say that if he really shared the allele (or alleles) responsible for the familial mood disorder it (or they) must be viewed as being nonpenetrant. Yet David once had a period of depression in response to a personal crisis and also on occasion displayed levels of vivacity that might have been hypomania, a mild form of mania. Is there, then, a formal way to call attention to the possibility that even though David's behavior doesn't justify his classification as manic-depressive his mood fluctuations might still be a manifestation of the same allele (or alleles) that had been passed on to Flora and Jerry?

Marked variability in the outward expression of alleles is in fact very common. Geneticists have a term for this—variable expressivity—that might well apply to David. In contrast with penetrance, an all-or-none property (like a light switch turned on or off), variable expressivity is a graded property (like a dimmer switch turned to one of many settings from very low to very high). And just as a retrospective (and somewhat controversial) genetic diagnosis applied to an English king gave us a memorable example of penetrance, so can a retrospective (and somewhat controversial) genetic diagnosis applied to an American president—Abraham Lincoln—give us a memorable example of variable expressivity.

The evidence of Lincoln's hereditary condition (presumably transmitted from his mother, Nancy Hanks) was his appearance, described by *The Times* of London as that of a "tall, lank, lean man, considerably over six feet in height, with stooping shoulders, long pendulous arms terminating in hands of extraordinary dimensions, which, however, were far exceeded in proportion by his feet." When these features are considered along with the often remarked upon looseness and mobility of Lincoln's joints, experts have concluded that our sixteenth president probably had Marfan's syndrome. This disorder, first described at the end of the nineteenth century (long after Lincoln's death) by a French physician, Antoine Marfan, is now known to be caused by a dominant allele of a gene that encodes a protein called fibrillin.

Like Lincoln, the typical person with Marfan's syndrome tends to be extremely tall and to have disproportionately long arms and legs and huge hands and feet. Those affected usually also have a variety of other problems, all—like their unusual body proportions—explainable by a flaw in their connective tissues produced because half their fibrillin is abnormal. Among the most common of these abnormalities is a defect in the wall of the aorta, the large artery that emerges directly from the heart, which may be so weakened that it will burst and cause death. But Lincoln, who was fifty-six when he was assassinated, showed no sign of this defect. So why do experts still consider it likely that he had Marfan's syndrome?

The answer advanced by these retrospective diagnosticians—and the reason I have been telling you these speculations about Lincoln—is that the expression of the manifestations of Marfan's syndrome varies greatly among those affected. Though everyone who inherits a defective allele shows some evidence of abnormality—making Marfan's syndrome 100 percent penetrant (the light switch is always on)—the degree and nature of the abnormalities vary enormously (the dimmer switch may be set at different levels). In some people there are skeletal malformations such as a twisted spine. In others, the skeletal abnormalities are minor, but the aortic weakness may give way, causing sudden death. Still others, like Lincoln, may have no more than an unusual appearance—a low setting on the dimmer switch.

Because variable expressivity, like non-penetrance, is common in hereditary disorders, it is reasonable to expect that it might be a feature of mood disorders, like those in Michael's family. If the mood disorders in that family are indeed influenced by critical alleles of one or more mood genes, Flora, Jerry, Max, and David might well share them but each express them differently—presumably because of individual differences in genetic background and various environmental factors. Even Michael's ups and downs might be a mild expression (like Lincoln's appearance, a low setting on the dimmer switch) of the same allele or alleles that he had presumably inherited from Flora and passed on to Jerry.

But isn't it premature to talk about penetrance and expressivity of mood gene alleles when we don't know for sure that they even exist? The reason to take such speculations seriously is that evidence is actually already available that can be most easily interpreted as reflecting the penetrance and expressivity of mood gene alleles. It was obtained by taking advantage of an experiment of nature: the existence of twins.

THE IDEA OF using twins to investigate the importance of heredity in human behavior was proposed in Mendel's lifetime by another major nineteenth century geneticist, Francis Galton. Although he was born in the same year as Mendel, 1822 (and three years after Garrod's father, Alfred, who became one of Galton's friends) his circumstances at birth could hardly have been more different. Whereas Mendel's parents were peasants, Galton was the son of a prosperous banker, grandson of the esteemed biologist Erasmus Darwin, and cousin of the illustrious Charles Darwin, whose 1859 publication, *The Origin of Species,* would revolutionize biology. And whereas Mendel was able to attend the University of Vienna only by a twist of fate, Galton went to Cambridge University with exalted expectations. Though these hopes were temporarily dashed when, struggling to earn an honors degree in mathematics, the twenty-year-old Galton had what has been called a "nervous breakdown," the substantial inheritance he was left by his father two years later freed him to be a gentleman-scholar like his cousin Charles.

Galton's approach to genetics was strikingly different from Mendel's. His subjects were people, not peas; the complex human traits he examined could not be reduced to single hereditary factors in the same way as the color of flowers or the texture of seeds; his data analysis was done with statistical methods that he helped to create rather than by simply computing a ratio. Fascinated by the work of Carl Friedrich Gauss, the German mathematician who devised the famous bell curve as a way of canceling out the effect of human errors on physical measurements (such as the distances of planets), Galton was the first to apply statistical methodology to the study of human variations themselves. From this work he became convinced of the enormous value of statistical techniques: "Their power of dealing with complicated phenomena is extraordinary. They are the only tools by which an opening can be cut through the formidable thicket of difficulties that bars the path of those who pursue the Science of man."

One complex human trait that lent itself well to Galton's approach was height. To evaluate the role of heredity Galton compared the heights of children with those of their parents, using a statistical method for measuring correlations that he had developed for this purpose. He was able not only to demonstrate quantitatively that tall parents tended to have children who were taller than average and that short parents tended to have children who were shorter than average, but also that children of a tall parent and a short parent tended to be of intermediate height. These results differed strikingly from Mendel's finding (not known to Galton) that pea plants were either tall or short, but never in between.

For many years after its rediscovery, Mendel's work seemed inapplicable to the complex human traits that Galton studied. The formal reconciliation of the Mendelian and Galtonian approaches came in 1918 through the theoretical work of Ronald Fisher, the second occupant of the professorship at University College, London, that Galton endowed in his will. Fisher showed that traits such as human height can be thought of as being influenced by the combined effects of alleles of multiple genes, each of which is transmitted in accordance with Mendel's laws. In fact, human height is a classic example of what is called not only a complex trait—one that reflects multiple environmental as well as genetic influences—but also a *quantitative* trait, with continuous variations over a wide range reflecting additive or interactive effects of alleles of many genes and environmental factors. Pea-plant height, on the other hand, is not only a monogenic trait but also a *qualitative* trait, with only two distinct phenotypes—tall plants or short plants.

Human height was not the only complex trait that interested Galton. In 1865, the same year that Mendel gave his famous lectures in Brno, Galton published two articles in which he expressed an interest in the hereditary basis of something much more controversial: human personality. Four years later, in his book *Hereditary Genius,* Galton described his studies of an aspect of personality that he called "eminence," a trait far more difficult to define than height. Nevertheless, starting with a particularly illustrious group, "the hundred most distinguished men," Galton concluded that the characteristic was familial: 26 percent of their fathers and 36 percent of their sons were "eminent," as were 8 percent of their grandfathers, 10 percent of their grandsons, and 2 percent of their great-grandsons.

Galton took this to mean that the familial tendency to "eminence" was, at least in part, hereditary. But he also recognized that the families in which "eminence" was concentrated were mostly privileged, raising the obvious possibility that favorable upbringing might play the major role ("when a parent has achieved great eminence, his son will be placed in a more favorable position for advancement than if he had been the son of an ordinary person"). To help evaluate the relative contributions of heredity and upbringing, Galton proposed using twins that were either "alike at birth" (that is, identical), or "unlike at birth" (that is, fraternal). And even though Galton never fully developed this approach, and never proposed to make use of yet another experiment of nature—identical twins separated at birth and raised apart—he helped set the study of twins in motion in a paper he published in 1876, "The History of Twins as a Criterion of the Relative Powers of Nature and Nurture." (Its title is derived from Shakespeare's *The Tempest*, in which Prospero gives his opinion of Caliban—"a born devil on whose nature nurture can never stick"—an opinion Galton probably shared.)

The main reason that twins have turned out to be such a godsend for behavioral geneticists is that they come in the two varieties that Galton had called attention to: identical (derived from the same fertilized egg, and thus having exactly the same alleles of every gene) and fraternal (each derived from a different egg, and thus having no more genetic similarity than any other two siblings). Assuming that the shared environments of sets of identical twins and sets of fraternal twins are roughly equal (which appears to be the case), comparing their degree of similarity gives an indication of the relative contributions of nature and nurture. On the one hand, if a pair of identical twins are more similar than a pair of fraternal twins in some phenotype, such as height, this observation argues for the importance of nature—that is, shared alleles—in controlling the phenotype in question. On the other hand, if a pair of identical twins have no greater similarity than a pair of fraternal twins in some phenotype such as the use of English as a primary language, this indicates that nurture is responsible.

Useful though this approach can be, it is no easy matter to assemble enough information about sufficient numbers of twins. But the method could be applied to the study of manic-depressive illness because of a Danish program that had been established early in the twentieth century to collect psychiatric data on twins. Using this data, codified as the Danish Psychiatric Twin Register, a group led by Aksel Bertelsen identified 69 pairs of identical twins and 54 pairs of same-sex fraternal twins, in which at least one of the twins ("Twin 1," the index case) had either bipolar disorder (episodes of both mania and depression) or unipolar disorder (only major depression). The other member of the pair ("Twin 2") was classified in one of four categories: bipolar; unipolar; "normal" (that is, with no signs of a mood disorder); and "partial." The last "partial" category comprised Twin 2s who showed some signs of a mood disorder but who didn't fit into the unipolar or bipolar categories.

A major finding of this study is that members of an identical twin pair showed a much greater similarity, or "concordance," than did members of a fraternal twin pair, supporting the importance of nature in the vulnerability to mood disorders, especially bipolar disorder. Of twin pairs in which the index case had bipolar disorder, 63 percent of identical twin pairs but only 9 percent of fraternal twin pairs were concordant. While supporting a genetic basis for bipolar disorder, the study also provides clear evidence for variable expressivity and non-penetrance of mood genes. Of the remaining identical twin pairs in which the index case had bipolar disorder, 34 percent showed evidence of variable expressivity, with 17 percent of Twin 2s having unipolar disorder and 17 percent classified as "partial." One (3 percent) of the 34 index cases with bipolar disorder had an identical twin who was "normal," indicating non-penetrance.

So the findings with twins increase the plausibility of the explanation we have been considering for the pattern of mood disorders in Michael's family. For even in those twins who are identical, their every allele and much environment in common, there are often different phenotypes in members of a pair. It is little wonder that Michael's particular combination of genetic and environmental backgrounds might have spared him from developing any symptoms, even while he would have to be carrying the critical alleles of mood genes in order to pass them from Flora to Jerry.

WHEN GARROD CONCLUDED that the transmission of AKU was consistent with Mendel's first law, he could rejoice in the realization that some features of human heredity are as easy to analyze as those that had been described in peas. But Garrod also came to understand that AKU was an unusually simple case—the product of alleles of a single causative gene. In his second classic book, *Inborn Factors in Disease*, published thirty years after his stunning application of Mendelian principles to disease and four years after he had retired as Regius Professor of Medicine at Oxford, Garrod turned his attention from hereditary causes of rare disorders to hereditary susceptibilities to common disorders. Having made his mark by working on a condition that was qualitative and Mendelian, Garrod had become interested in conditions that were quantitative and Galtonian.

Though Garrod could deal with hereditary susceptibility only in a very general way, we now know the precise structure of alleles of some of the susceptibility genes involved in certain common diseases such as high blood pressure or diabetes. To make the great leap from a simple case like AKU to the actual identification of alleles involved in such complex disorders required the development of a revolutionary approach to genetic analysis to which we will now turn—an approach that is being applied to manic-depressive illness. The reason this approach has proved to be so powerful is that it goes directly to the very heart of genetics—to the chromosomes in which genes reside, and to the DNA of which they are made.

5

SOME TOOLS
FOR THE HUNT

That the fundamental aspects of heredity should have turned out to be so extraordinarily simple, supports us in the hope that nature may, after all, be entirely approachable. Her much-advertised inscrutability has once more been found to be an illusion due to our ignorance.

—Thomas Hunt Morgan

Gregor Mendel, Archibald Garrod, and Francis Galton, the three main characters in the story of medical genetics whose contributions I have described so far, lived in that glorious period when great discoveries could be made by amateur scientists working alone and in their spare time. A priest who discovered the basic principles of heredity while spending his summers breeding peas. A physician with a busy practice and an interest in the chemical composition of urine who applied those principles to human diseases. A gentleman-scholar with a passion for counting who developed statistical techniques for estimating the extent to which heredity contributes to complex human traits. Each of them brilliant. All of them amateurs.

Once this foundation was laid the next steps required a new breed of scientists: professionals in universities, with access to students, assistants, and specialized instruments. Among them was another great pioneer, Linus Pauling, once described as "the most brilliantly versatile and productive physical chemist of the [twentieth] century; more than a scientist, a force of nature." Having spent many years studying the chemical properties of proteins, in the 1940s Pauling became interested in applying his theories to an understanding of inherited human diseases—theories that would ultimately guide the hunt for mood genes. Garrod had already grasped the general idea that inherited diseases were reflections of protein abnormalities; now Pauling was eager to determine their exact chemical nature. His opportunity came when he learned about an interesting feature of a hereditary disorder called sickle cell anemia, which is common among people of African descent.

Sickle cell anemia gets its name from a property of the red blood cells of those affected—the tendency of the cells to switch from being round to being shaped like sickles. With this abnormality comes diminished flexibility of the red blood cells, which causes them to break as they are pumped through the body's small blood vessels, resulting in anemia. The sickle-shaped cells can also clog these small blood vessels and cause attacks of severe pain, called sickle cell crises. The clogging can, in turn, damage many organs over time, leading to a wide range of symptoms that may vary greatly from person to person.

Pauling's curiosity about sickle cell anemia can be traced to his interest in the shapes of particular proteins. Among them was hemoglobin, the major protein of red blood cells, which carries oxygen throughout the body. In the course of a dinner meeting in February 1945 of a small committee charged with modernizing medical research in the United States, William Castle, a physician who specialized in blood diseases, told Pauling that sickling occurred only in veins and capillaries (which have low levels of oxygen), but not in arteries (which are rich in oxygen). At the time neither Castle nor Pauling had any idea that all the manifestations of this disease could be traced to hemoglobin. But, for Pauling, Castle's news triggered an elaborate hunch: that sickle cell anemia was completely explainable by *an inherited defect in hemoglobin* that forces it to change its shape in conditions of low oxygen.

Pauling was encouraged to pursue his hunch because hemoglobin was readily available in large amounts and so was a convenient object for research on the relationship between the chemical structure of a protein and its shape. Pauling and his students set out to determine if they could distinguish the chemical structure of the hemoglobin in the red blood cells of people with sickle cell disease (called hemoglobin S, for sickle) from that of the hemoglobin in the red blood cells of normal adults (called hemoglobin A, for adult). Using a simple technique called electrophoresis, in which a solution of hemoglobin is placed in a piece of glass tubing with a positive electrode at one end and a negative electrode at the other (so that an electric current can be passed through the solution), one of Pauling's students, Harvey Itano, soon found that the two forms of hemoglobin were easy to tell apart: the A form, which was more negatively charged, moved closer to the positive electrode. Furthermore, in people with one copy of the normal allele and one copy of the sickle cell allele, half the hemoglobin was A (that is, it moved to the A position in the electric field) and half was S (it moved to the S position in the electric field), indicating that each allele was directly responsible for the manufacture of its share of this protein. Unlike Mendel and Garrod, who had to infer the presence of alleles from phenotypes such as flower color or urinary pigment, Pauling and Itano *could see the direct reflections of the alleles* because of different physicochemical properties of the specific protein molecule whose construction each allele encoded.

But what was the precise chemical basis of the difference between hemoglobin S and hemoglobin A that gave the latter its greater negativity? The answer was eventually provided by Vernon Ingram, who was working at the Cavendish Laboratory in Cambridge, England, where James Watson and Francis Crick had in 1953 determined the structure of DNA. Encouraged by Crick, who was at the time pondering the relationship between DNA and proteins (that would lead him to the Central Dogma), Ingram decided in 1955 to try to determine the exact chemical difference between normal hemoglobin and sickle cell hemoglobin by applying new techniques for protein analysis that had been developed by Frederick Sanger, also at Cambridge. With these techniques the samples of normal and sickle cell hemoglobin were each first treated with an enzyme that chopped them into specific fragments. After the fragments were separated by electrophoresis (but on blotting paper instead of in a tube, and in combination with another procedure that separated the fragments on the basis of a different physicochemical property), the single small fragment that was different in hemoglobin A and hemoglobin S was removed from the paper. Then the distinguishing fragments from A and from S were compared, using methods that determined which amino acids each contained and how they were arranged.

By 1957 Ingram was able to report that the difference between the normal hemoglobin A and the sickling hemoglobin S was a single amino acid substitution in the component of this complex molecule called beta-globin. In normal beta-globin the sixth amino acid in the protein chain is glutamic acid. In contrast, the beta-globin from people with sickle cell anemia has a different amino acid, valine, in the sixth position. Because glutamic acid is negatively charged and valine is not, the different movement of hemoglobin A and hemoglobin S in the electric field is accounted for. Furthermore, when this single valine replaces glutamic acid, the protein is sufficiently affected that in low oxygen it undergoes a shape change. As Pauling had suggested, a specific chemical change in this protein is what causes sickling, clogging of tiny blood vessels, destruction of red blood cells, and all the resultant symptoms—*the first detailed molecular explanation for a disease!*

Ingram's finding also suggested the reason for this difference: a structural change in the gene encoding beta-globin that had converted a normal (S) allele that produced normal hemoglobin A into a sickle cell (s) allele that produced the defective hemoglobin S. Presumably this came about because the S allele in the egg or sperm of an ancestor had been damaged (perhaps by a toxic chemical), creating a mutation—a change in the DNA structure—that resulted in the s allele that was transmitted to descendants. We now know that this specific change in the DNA is exactly what must have happened, because the structure of both the S allele and the s allele have been identified by modern chemical techniques. As expected, in people with sickle cell anemia the sixth codon in the gene for beta-globin is an instruction for valine instead of glutamic acid. It is just this sort of variation in the structure of mood genes that may alter the susceptibility to manic-depressive illness.

PAULING'S HUNCH THAT sickle cell anemia might be due to an abnormality in the structure of hemoglobin built on a large body of knowledge about this protein. To hunt for the genetic basis for manic-depressive illness, some scientists are trying to use a similar approach called the "candidate-gene approach," or "round up the usual suspects." Just as Pauling was led to hemoglobin because of an established property—its known link to oxygen—the candidate-gene approach begins by considering the properties of already known brain proteins and the genes that encode them. The idea is to look over these properties in order to make some educated guesses about likely "candidates," or "suspects," which, if mutated, might affect mood.

Among these suspects are the many proteins that participate in the manufacture or inactivation of serotonin, norepinephrine, and dopamine, three brain chemicals (called neurotransmitters) known to be affected by drugs that are used to treat mood disorders. One protein that has attracted particular attention is the serotonin transporter, a protein that inactivates serotonin and that is the target for antidepressant drugs such as Prozac. Following the line of thinking in Pauling's and Ingram's work on hemoglobin, researchers have scrutinized the gene that encodes the serotonin transporter for mutations in people with mood disorders. Another popular candidate has been the gene that encodes an enzyme called tyrosine hydroxylase, which participates in the manufacture of dopamine and norepinephrine. But so far no luck. Although the candidate-gene approach remains very promising, and has helped find genes involved in other diseases, it has not yet succeeded in finding a mood gene.

A major limitation of the candidate gene approach is that most of the hundred thousand human genes have not yet been identified, and the functions of many of those that have been are still unknown. So until we learn a great deal more about our genetic repertoire, many candidates will be overlooked, since the basis for their candidacy will not yet have been discovered. The other limitation of this approach is that it depends on clever hunches as to the most plausible genes responsible, hunches that we can't really rely on. If asked about the genetic basis for wrinkled peas, who would have guessed that it was a gene that converts sugar to starch?

Fortunately, there is another, more systematic way of uncovering relationships between genes and a disease such as manic-depression—a way that is very different from the candidate-gene approach. Called the linkage approach, it seeks to identify suspects in a manner that may, at first, seem strange and irrelevant: by looking for a close physical proximity, or "linkage," between the unidentified gene that is involved in the disease and a known gene called a marker. What makes this approach possible is that every gene, whether already known or *yet to be identified,* has a specific location in our DNA: a specific address. And just as our home addresses can be pointed out, not only by street numbers but also by proximity to neighbors ("I don't know L's address but she lives somewhere between J and P"), so too can the locations of genes be pinpointed. Then, once the location of a gene is known, other techniques—modern versions of Ingram's, based on DNA chemistry rather than protein chemistry—can be used to scrutinize the gene residing at this location. By comparing the precise structure of this gene in normal people with that in people with a particular disorder, evidence of a mutation may be found (as in Ingram's case) to help clinch the identification.

The best (though imperfect) analogy I can offer of a gene hunt based on linkage comes from World War II movies in which the enemy spy with the secret radio transmitter is tracked down by gradually homing in on the source of the radio waves. For just as the good guys with the signal detector eventually pinpoint the house—nabbing the spy-whom-no-one-suspected—so too can the linkage approach find *unknown* or *unsuspected* genes that play a role in various diseases, *purely on the basis of their locations.*

And just as the radio signal from the spy is plotted first by region in the city, then to an address on street maps, so too is the signal from the unknown allele—the trait it influences—plotted first to a region called a chromosome, then to a zip code, and finally to a unique address on a genetic map. No assumptions, guesses, or hunches about the identity of this suspect are required. Only when the search is completed are other means of final identification brought to bear (such as fingerprinting the spy or chemical analysis of the alleles of the gene to find the precise mutation). And because the genetic linkage method uses a technology that is as systematic and powerful as that of the radio–signal-detection method, it too has a very good chance of succeeding in the end.

THE LINKAGE APPROACH, which is now being used to hunt for many genes, including mood genes, builds on work done early in this century by Thomas Hunt Morgan, the first Nobel laureate in genetics and one of the founders of this field in America. Born in Kentucky in the aftermath of the Civil War to a prominent Southern family (its members included Francis Scott Key, who wrote "The Star Spangled Banner," and Brigadier General John Hunt Morgan, the famous cavalry raider known as the Thunderbolt of the Confederacy), Morgan migrated to New York City in 1904 to join the zoology faculty of Columbia University. A few years later he became interested in the use of the common fruit fly, *Drosophila melanogaster,* to study the principles of heredity, attracted in part to flies because he could lure endless subjects to his laboratory simply by placing ripe bananas on the windowsill.

A frugal man, Morgan was also attracted to the tiny size of these creatures, which he housed in half-pint milk bottles that he appropriated from the cafeteria (no need for a garden) and which he could feed inexpensively with bits of fermented banana. Among other advantages of fruit flies for genetic studies is their rapid development from fertilized eggs to reproductively active adults. This maturation process takes only about twelve days (much faster than peas), enabling researchers to study inheritance over many generations in a short period. And because a single mating can provide several hundred fertilized eggs, it is easy to obtain a virtually limitless supply of subjects for genetic analysis.

Despite these attractive features, Morgan faced a major problem. Studies of heredity require readily observable phenotypes like the variations in flower color or seed shape that Mendel worked with (or a human disease such as manic-depression). Because farmers had been breeding peas for centuries, Mendel had lots of possibilities to choose from; but Morgan had no distinctive phenotypes to work with. What he needed were mutant flies with genetic abnormalities that were reflected in some aspect of their appearance. And the only way to find them was to examine flies one at a time, looking for a novel feature. To this end he and his assistants anesthetized flies with ether and inspected each one of them with a jeweler's magnifying glass, hoping that something interesting would turn up.

Finally, in May 1910, after many months of searching, Morgan discovered a male fly that seemed so precious he took it home that night to make sure it wouldn't get lost: this fly had white eyes. Because normal fruit flies have red eyes, white eyes would be as easy to track as white or purple flowers. Furthermore, by breeding the white-eyed male with a red-eyed female (according to Morgan's biographers, this mutant male "mustered enough strength to mate with a normal female before dying"), Morgan quickly established that red is dominant, because there were only red-eyed offspring in the first generation. And in the next generation (by brother-sister mating, as in Mendel's studies) the progeny were in the expected Mendelian ratio of three red-eyed for every one white-eyed. So far, no surprises.

But on closer inspection, there *was* something baffling. Instead of equal numbers of white-eyed males and females, as one might expect, all the white-eyed flies from the brother-sister mating were *males*. Did this mean that femaleness precluded development of white eyes? Certainly not, since other breeding experiments eventually led to the appearance of white-eyed females. What explained their absence after the original brother-sister mating?

In answering the question Morgan made two critical inferences: genes must reside on chromosomes (the "colored bodies" observed in the cell nucleus that got their name because they so avidly absorbed certain stains); and each gene must reside on a particular chromosome. Soon he was to make a third inference that incorporated the first two: that each gene must reside at a *specific place* on a particular chromosome. This profoundly important inference eventually provided a conceptual basis for hunting for genes through linkage analysis, whether they affected the color of flies' eyes or the mood disorders of human beings.

A major reason that Morgan could make such far-reaching inferences from such limited experimental data was that he knew about the existence of chromosomes. These structures had already been seen under the microscope several decades earlier, but nobody had figured out their function. It *was* known that all chromosomes existed in pairs in females but that in males one chromosome was unpaired—a difference that proved to be the basis for sex-linked inheritance. By 1905 the chromosome that was paired in females but not in males had been named the X chromosome, and the much smaller unpaired chromosome found only in males had been named the Y chromosome. It had also been shown that (unlike human cells, with forty-six chromosomes) each fruit fly cell, with the exception of eggs or sperm, has a total of eight chromosomes: three pairs (called autosomes and numbered 1, 2, and 3) and two others called sex chromosomes—two Xs in females and, in males, an X and a Y.

Given this information, Morgan concluded that his experimental results could be explained if the eye-color gene had a specific location on the X chromosome, which is unpaired in the males; and if there was a normal dominant allele, R, that gave red eyes and a mutant recessive allele, r, that gave white eyes. In the absence of the mutation all flies have only R alleles, two copies in females (since females have two X chromosomes), but only one copy in males (since males have only one X chromosome).

Now consider the original white-eyed mutant that Morgan had chanced upon, some of whose descendants are displayed in the diagram on the following page. Being male, with only one

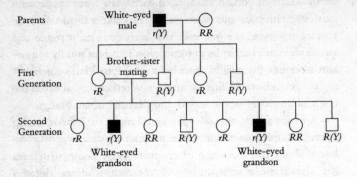

X chromosome, the only eye-color allele the mutant fly had was *r*; and lacking *R*, he had white eyes. Yet when this mutant male was bred with red-eyed females with genotype *RR*, all his off-spring had red eyes. His daughters had the genotype *rR*, because they received both his X chromosome (with its *r* allele), and the mother's X chromosome (with its *R* allele); and his sons had the genotype *R*, because they received his Y (which both made them male and lacked an eye-color allele), and a maternal X (with its *R* allele). Only in the next generation could the mutation show up, in his daughters' sons—because the daughters had an *r* allele to pass on to some of their male offspring.

Once this pattern of inheritance was understood, it did not take long to find many more mutations in fruit flies, such as a yellowish body color instead of the normal brownish body color, that were also transmitted in this sex-linked way, indicating that the relevant genes were also on the X chromosome. But even more interesting was the association *between* the sex-linked genes—such as between the gene that controlled red vs white eye color and the gene that controlled yellow vs brown body color. Though specific alleles might start out on the same X chromosome, they didn't always end up that way.

For example, when Morgan mated a mutant male (with a yellow body and white eyes) and a normal female (with a brown body and red eyes) and then mated their offspring among themselves, there was a big surprise. Although many descendants of the brother-sister matings had both white eyes and yellow bodies, a few descendants had only one or the other of the mutant phenotypes (that is, white eyes and a brown body or red eyes and a yellow body). The ancestral male had passed on the white-eye allele and the yellow-body allele that had both resided on the same (and only) copy of his X chromosome; but *these two alleles did not always remain together on the same chromosome in subsequent generations.* In some cases they had split apart and come to reside on separate X chromosomes in separate descendants. And it is precisely such separations (on all chromosomes, not just X), and the frequency with which they occur in descendants—that underlie the development of the gene maps that are now being used in hunting for mood genes.

So WHAT IS the basis of this occasional split of the eye-color allele and the body-color allele that were once on the same ancestral X chromosome? Again the explanation comes from microscopic studies of cells, specifically the behavior of chromosomes during sex-cell formation. In contrast with usual cell division, in which all the chromosomes are copied and then distributed to daughter cells (as goes on continuously in many body tissues), the cell division that gives rise to sex cells is more complex. As sex cells are formed, they wind up with *only one member of each parental chromosome pair.* A fruit fly's body cells have a total of eight chromosomes (three pairs of autosomes and two sex chromosomes), but its sex cells have only four chromosomes, one each of chromosomes 1, 2, and 3 (instead of three full pairs) and either an X or Y (instead of both). This will be each parent's contribution to a particular offspring, who must, of course, wind up with eight chromosomes (by the fusion of two sex cells, one from each parent). Now comes the critical new point: the distribution of chromosomes to sex cells occurs in a manner that accounts for the occasional separation of the yellow-body allele and the white-eye allele that Morgan had observed.

How does this work? Consider a female fly with eight chromosomes, four from its father (1F, 2F, 3F, and XF) and four from its mother (1M, 2M, 3M, and XM), each with many different genes. As the fly manufactures eggs, each egg winds up with four chromosomes. But which set? The F set or the M set?

The answer is that sex cells get different mixtures of F and M, which helps to assure diversity of offspring, a potential evolutionary advantage. Substantive mixing could be achieved by randomly selecting, on the average, two F chromosomes and two M chromosomes, so that a sex cell would wind up with some mixture. Were this scheme followed, each parent could generate sixteen different sex cells, such as (1M, 2F, 3M, XF), (1M, 2F, 3F, XF), (1F, 2F, 3M, XM), and so forth. Because eggs and sperm combine randomly to produce embryos, these conditions would provide even monogamous parents with a wide variety of potential offspring.

But nature has not taken this easy way. Instead, given the potential value of diversity for natural selection, an additional process has made the variety of possible offspring of a mating even larger. In this process, portions of matched chromosomes (such as 1M and 1F) are *exchanged* before being distributed to the sex cells. The process, called recombination, or crossing over— which affects every single chromosome during sex-cell formation—is what accounts for the occasional separation of the yellow-body allele and the white-eye allele that had surprised Morgan. Crossing over is like cutting a deck of cards before dealing: during sex-cell formation there is always at least one cut in each chromosome, followed by crossing over.

Crossing over begins after the paternal and maternal copies of an individual chromosome have replicated. Then, rather than immediately being distributed to sex cells, the replicated chromosomes come together as pairs, with maternal and paternal copies lined up next to each other. To get a simplified picture of what happens, let us consider a copy of chromosome 1 derived from father fly and a copy of chromosome 1 derived from mother fly, each arbitrarily divided into ten segments (Ma–Mf and Fa–Ff), which line up as shown at the top of the facing page:

Chromosome 1 from mother

| Ma | Mb | Mc | Md | Me | Mf | Mg | Mh | Mi | Mj |

Chromosome 1 from father

| Fa | Fb | Fc | Fd | Fe | Ff | Fg | Fh | Fi | Fj |

In the next step, the F chromosome and the M chromosome both break. The site of the break is random—that is, it can be anywhere along the length of a chromosome, just as cutting a deck of cards can be done at any position in the deck—but it must be at the same point on both members of the pair (for example, between b and c). Then the pieces are exchanged, crossing over from one chromosome to the other, and are zipped back together again to make intact new versions. So, if the break was between b and c, the two chromosomes that result are:

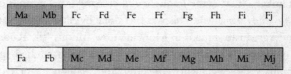

| Ma | Mb | Fc | Fd | Fe | Ff | Fg | Fh | Fi | Fj |

| Fa | Fb | Mc | Md | Me | Mf | Mg | Mh | Mi | Mj |

Two versions of chromosome 1 after crossing over. Each is then packaged in a different sex cell.

The outcome of this process is that the version of chromosome 1 that is transferred to a particular egg or sperm is *partly* derived from the father and *partly* from the mother. Because the site of the break is random, the contributions from F and M vary from sex cell to sex cell, increasing the potential diversity of offspring. Since all the chromosome pairs exchange F and M pieces, the potential diversity becomes enormous.

What Morgan realized as he accumulated more and more mutants is that genes that are close together on a chromosome (for example, one at b and the other at c) tend to stay together during crossing over (just as cards that happen to be next to each other in a deck are unlikely to be separated by a random cut), whereas genes at opposite ends of a chromosome (that is, at a and j) would always be separated by crossing over (just as a cut always occurs between the card on the top of the deck and the card on the bottom of the deck). Most genes are on different chromosomes or are far apart on the same chromosome, explaining Mendel's finding of independent assortments of pairs of traits such as seed shape and color. But some genes, such as the genes that control red or white eye color and yellow or brown body color of flies, happen to be so close together on the same chromosome that cuts between them are very infrequent. Such genes are said to be *linked*. And the closer they are on the same chromosome, the less likely they are to be separated by a random cut, and the more strongly they are linked—that is, the more often the phenotypes these genes affect will be transmitted *together* from a particular parent to a child.

In little more than a year after the discovery of the first white-eyed fly, enough mutations on the X chromosome had been found to make it possible to begin to express the strength of linkage (that is, the frequency of crossing over) in units of *distance on a chromosome*. This was first done by Alfred Sturtevant, an undergraduate student in Morgan's now famous "Fly Room" in Columbia's Schermerhorn Hall (where I too studied undergraduate biology), who converted the frequencies of crossing over between pairs of genes on the same chromosome (such as the red-or-white-eye-color gene and the yellow-or-brown-body-color gene) into a graphical depiction of the distance between these genes—the first genetic map. Since all alleles of a gene occupy the same location on a chromosome, a convenient term to refer to a gene and all its alleles is "locus" (plural, "loci"—from the same root as "location"). Geneticists often prefer to use "locus" rather than "gene" when they have mapped the factor that controls an inherited trait to a specific location on a chromosome but have not actually isolated the gene as a bit of DNA, and may have no idea what protein it encodes. In these cases, all they know is that the alleles that control alternative forms of that trait, such as yellow or brown body color, reside at a specific location. Though the identity of the spy has not been determined, the address of the radio signal has been found.

In flies, mapping loci is relatively easily done, by examining and tabulating the phenotypes of thousands of offspring of appropriate matings. For example, in the case of white eyes and yellow bodies in flies, Sturtevant (working only with a magnifying glass and a notebook) found that the separation between these two mutant phenotypes was very unusual, occurring in only about fifteen out of a thousand sex cells that were formed (as estimated by the phenotypes of the flies that resulted). Other mutant phenotypes, such as rudimentary wings and miniature wings, also known to be controlled by genes on the X chromosome (because they too showed sex-linked patterns of inheritance) separated more frequently. Eventually the results of this mapping were expressed in terms of a unit of distance, now called a centimorgan, a term proposed by the great British geneticist, J.B.S. Haldane, in honor of Morgan, and abbreviated cM.

A centimorgan is defined as the distance between two loci that are separated by crossing over only once in the formation of every hundred sex cells. So, for example, since Sturtevant found fifteen recombinations per thousand offspring between the white-eye locus and the yellow-body locus (or 1.5 per hundred), the distance between these two loci would be described as 1.5 centimorgans. With data of this sort, Sturtevant was able to plot the relative locations of the handful of then known loci on the first map of the X chromosome. The diagram on the following page shows the locations of four of them from a contemporary map on which the yellow-body locus is at one end of the chromosome, the white-eye locus is 1.5 centimorgans away, and the miniature-wing locus and rudimentary-wing locus are, respectively, 34.6 centimorgans and 53 centimorgans from the white-eye locus. And just as Sturtevant used already known loci as markers on the fruit fly map to determine the newly discovered loci, so too can already known loci on the human map be used as markers to determine the addresses of loci that influence phenotypes such as manic-depressive illness.

To illustrate how the linkage approach can be used to hunt for mood genes, it is easiest to start with a simplified example.

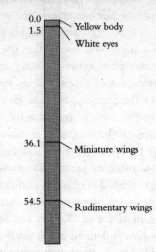

MAP OF THE FRUIT FLY X CHROMOSOME
WITH FOUR MARKERS USED BY STURTEVANT

0.0
1.5 — Yellow body
— White eyes

36.1 — Miniature wings

54.5 — Rudimentary wings

Imagine a large family with several dozen members with manic-depressive illness distributed over several generations. For the sake of argument, let us also suppose that the pattern of transmission of the mood disorder in this family is that of a Mendelian dominant disorder (such as Huntington's disease); and that the dominant allele, *M*, is fully expressed, so that everyone who inherits this allele winds up with manic episodes and other symptoms of the disease. Of course we already know that this situation is unrealistic: manic-depression is not a simple Mendelian trait. But I have conjured up this simplified example to illustrate how the linkage approach might be used to find *M*, because it can also be applied to more complicated problems.

Now let me add an additional feature to our hypothetical family: that those members with manic-depression all have brown eyes, reflecting a dominant *B* allele of an eye-color gene that will serve as the genetic marker in this case. To maximize the usefulness of this new condition for our purpose, let's also say that brown eyes are rare in the population we are considering (imagine it as being largely made up of blue-eyed Northern

Europeans), so that all the people who marry into this family are blue-eyed (genotype *bb*). And none of these blue-eyed spouses has manic-depressive illness.

Given these conditions, a linkage study might be initiated by a medical researcher interested in manic-depression who on meeting a member of this family (the index case) learns that many other family members also appear to be affected. In interviewing the patient the researcher might also notice various features, such as brown eyes and red hair (also rare in this hypothetical population and, for our purposes, also assumed to be transmitted as a Mendelian dominant). Gaining the cooperation of the family the researcher would then carefully examine all its members, and might find nine others with clear evidence of multiple manic episodes and other signs of manic-depressive illness. In interviewing these affected relatives he would notice that some are red-headed, but most are not, precluding the possibility that the gene that controls red hair is linked to *M*.

But eye color would turn out to be another matter. To his great delight, the researcher would find that all the patients with manic-depressive illness had brown eyes, and all those unaffected had blue eyes. The results of this hypothetical study are shown in the diagram on the following page, with filled-in symbols indicating people who have the mood disorder, and the eye-color gene alleles and mood-gene alleles shown below each symbol.

So what is the meaning of this hypothetical association between brown eyes and manic-depression? The association raises the possibility that just as with the white-eye and the yellow-body loci in Morgan's and Sturtevant's flies, the eye-color and the mood-disorder loci in human beings may be so close together on the same chromosome that there was no crossing over between them in the ten people with manic-depressive illness in this family.

It's also possible, however, that this apparent relationship between brown eyes and the mood disorder was only a chance association, just as flipping ten heads in a row can occur by chance. Put differently, the gene that controls eye color and the gene that influences the development of a mood disorder could

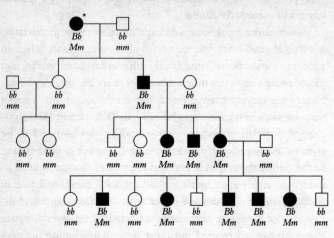

LINKAGE OF BROWN EYES AND MANIC-DEPRESSION
(HYPOTHETICAL EXAMPLE)

●, ■ Bipolar disorder

* Great-grandmother with bipolar disorder
B Brown-eye allele
b Blue-eye allele
M Mood-gene allele giving rise to manic-depression
m "Normal" mood-gene allele

actually be on different chromosomes, and their seeming linkage just a coincidence. To estimate the probability that an apparent linkage might occur by chance, statistical methods have been developed, the most commonly used of which is called a lod score ("lod" taken from "logarithm of the odds"). The lod score in this hypothetical case would indicate that the odds are better than 20 to 1 that the apparent linkage between the mood-gene locus and the eye-color locus did not occur by chance: pretty good odds, and the usual criterion for what is called "a statistically significant finding." But not proof. The only way to prove that the relevant locus has been found is to identify the gene and demonstrate the mutation, the change in its chemical structure. Finding evidence for linkage is only a step toward this ultimate goal.

BUT THIS IS just a fabricated example. It would require amazingly good luck to stumble upon an obvious inherited feature, such as eye color, that was controlled by a gene that just happened to sit so near to the locus of interest (in this case the locus of a mood gene) that linkage could be established. To use this approach in a systematic manner, thereby reducing the reliance on luck, it is necessary to have markers—known gene loci—distributed at regular intervals on each human chromosome, because the locus of interest could be *anywhere*. Establishing the locations of these markers on human chromosomes would generate a human genetic map, in the same way that Sturtevant created a genetic map for flies. Were the intervals between markers on the human genetic map sufficiently small, there would always be some markers that would be close enough to the locus of interest to show linkage.

This sounds fine in principle; but extension of the mapping approach developed for flies to the twenty-four different human chromosomes—twenty-two autosomes and the X and Y sex chromosomes—presented a formidable challenge. As with flies, in human beings all the early work was done with sex-linked traits, such as color blindness and hemophilia, because their distinctive pattern of sex-linked inheritance made it possible to assign their loci to the X chromosome. But determining their relative locations on the X chromosome was exceedingly difficult because it depended (as correspondingly with flies) on finding people with *both* color blindness and hemophilia, and studying the distribution of these traits in their offspring. And that is exactly the way these two loci were eventually mapped.

With the other human chromosomes there was, for many years, no progress at all. For in contrast with flies, in which new mutations could be deliberately generated by x-rays or chemicals, and in which individuals with particular combinations of mutations could be obtained by selective breeding, there were very few human phenotypes—"experiments of nature"—that lent themselves to such analysis. By 1940, Sturtevant, who had moved with

Morgan to Caltech twelve years earlier and who himself had become a major figure in genetics, began to despair that he would ever live to see even a crude map of human chromosomes.

Eager to find markers for a human genetic map, Sturtevant even considered using outlandish traits such as tongue rolling (sticking out the tongue as a rolled tube), which most people can do, but which others (like me) cannot. He was so serious about this that he published a paper in 1940 titled "A New Inherited Character in Man," in which he justified the possible use of this trait as a marker by proposing that the ability to roll the tongue was controlled by the dominant allele of a single gene. His idea was that the ability to roll the tongue reflected this dominant allele, and that the inability to roll the tongue reflected a recessive allele (just as red eyes in flies reflect one allele and white eyes another). And his hope was, presumably, to relate tongue rolling to other traits, in the same way that he had related eye color to body color in flies—making the "tongue-rolling locus" a marker on a human genetic map.

But, much to his regret, it turned out that Sturtevant had not been sufficiently rigorous in analyzing his tongue-rolling results; and his conclusion that this trait was controlled by a single gene turned out to be wrong (it seems, instead, to be polygenic and modifiable by learning). Sturtevant frankly acknowledged his error in 1965 in *A History of Genetics*:

> It is important that suspected cases of Mendelian inheritance in man should be recorded, so that they may be . . . incorporated in studies of possible linkage. . . . There is an unfortunate tendency, however, to accept cases as established when *the evidence is so weak that it would not be considered conclusive for any organism other than man* [italics added]. My own experience in the field may be cited as an example. About 70 percent of people of European ancestry are able to roll up the lateral edges of the tongue, while the remaining 30 percent are unable to do so. In 1940 I suggested that this difference is due to a pair of genes (the ability being dominant), though it was clear that a few people were able to learn to do it and that

there were a few discordant pedigrees. In 1952 Matlock showed . . . that even if an inherited component is an actuality (which is not certain), there is sufficient nongenetic influence to make the character practically useless as a genetic marker. But I am still embarrassed to see it listed in some current works as an established Mendelian case.

Sturtevant's grasp at a "tongue-rolling gene" is indicative of the desperation among those interested in human genetic mapping at the middle of the twentieth century and the seeming hopelessness of the task. For aside from a few markers that were known to be located on the X chromosome, the rest of the human genome remained uncharted territory. And even though mapping of human autosomes did finally get under way in the 1970s, progress was extremely slow. Little wonder, then, that during this long period hardly any attention was being paid to Kraepelin's proposal that there was an inherited vulnerability to manic-depressive illness, thus leaving the field to Freud's disciples. At the time that I first met Michael, in 1978, there still seemed to be no practical way of constructing a human genetic map of sufficient detail to do any sort of linkage work on manic-depressive illness, let alone guide a hunt for mood genes. There just weren't enough markers.

But as Michael and I were soon to observe with both awe and exultation, a truly major breakthrough was in the making. It was heralded by a revolutionary approach to the creation of a human genetic map that was in turn guided by an explosion of knowledge about the detailed structure of human DNA. The revolution began in the late 1970s with the realization that, just as people or flies have variations in their DNA that are reflected in readily observable differences, so too do they have other *hidden* variations that can be detected *only by direct DNA examination*—variations that can also serve as markers.

As this revolution has unfolded, thousands of these DNA-variation markers have been localized to specific sites on human chromosomes, providing a map of a precision and refinement far exceeding anything Morgan or Sturtevant could have imagined.

Furthermore, markers based on structural variations in human DNA turned out to be easy to measure, initially by adapting the electrophoresis procedures used by Pauling and Ingram to examine beta-globin, and now in even easier ways. All that is needed is the DNA from the cells contained in a few drops of blood and the proper biochemical reagents for each of the markers. No need to look at eyes or ask for tongues to be stuck out and rolled. This is *one-stop-testing*, with all the measurements made by the same basic procedure. And because this procedure can be systematically applied to all regions of all human chromosomes, it can in principle lead us to loci that influence any disease, including manic-depressive illness.

But how is this possible? How can there be variations in our genetic material that can serve as markers, yet produce no observable effects—no phenotypes? And how can we find these variations and put them to use? For answers we need to take a closer look at the details of the structure of human DNA.

WHEN IT WAS first shown that DNA is the genetic material, and that it functions by encoding proteins, it was assumed that genes were packed next to each other, without spacing, and that every single base pair was critical. But we now know—amazing though it may seem—that only a small fraction of human DNA encodes proteins, and that large regions of DNA within and between genes can undergo mutation with no outwardly observable effects. Nevertheless, mutations in such regions that cropped up in ancestors are transmitted to descendants just as reliably as mutations in other parts of the DNA, such as the mutation in the beta-globin gene that causes sickle cell anemia. We all carry these silent DNA variations, just as we carry other variations that determine our different phenotypes.

Such detailed knowledge of gene structure became possible only with the development of a series of techniques for studying DNA, beginning with the 1970 discovery of a group of bacterial proteins called restriction enzymes. Evolved by bacteria as protection against DNA-containing viruses, the different restriction enzymes bind to specific sequences of DNA base pairs and clip both strands of the double helix within these sequences. The benefit to bacteria is that invading viruses are cleaved and rendered non-infectious. The benefit to geneticists is that these enzymes allow for the cleavage of DNA (including human DNA), into precisely defined bite-size pieces.

The sites of the bites are the sequences of base pairs recognized by the particular restriction enzymes. For just as the shape assumed by an enzyme in Mendel's peas recognizes and fits into the shape of a precisely defined form of sugar (to convert it to starch), so does the shape of a restriction enzyme in a particular bacterium recognize and fit into the shape of a precisely defined sequence of base pairs in DNA (to cleave them). Since there are many such sequences at precise points along a double-strand of DNA, the bites give pieces of specific lengths, usually a few hundred to a few thousand base pairs long. For example, a restriction enzyme called *HpaI,* derived from a bacterial species called *Haemophilus parainfluenzae*, cleaves in the middle of the following double-stranded sequence:

$$\text{GTT} \mid \text{AAC}$$
$$\text{CAA} \mid \text{TTG}$$

The discovery of restriction enzymes was a major step in the DNA revolution. One offshoot was Yuet Wai Kan and Andrée Dozy's 1978 proposal, made while they were comparing the DNA structure of the normal beta-globin gene with the gene in people with sickle cell anemia, that *the sites themselves* could be used as milestones in a genetic map, because many of them have undergone easily detectable mutations that are carried by some people but not others. As an example, consider a mutation that causes a substitution of a G-C pair for an A-T pair in the *HpaI* site, to give the following:

$$\text{GTCAAC}$$
$$\text{CAGTTG}$$

The remarkable consequence of such a mutation is that *HpaI* does not recognize this altered sequence and *no longer cleaves* the DNA at this site, so that a particular fragment is no longer formed.

Since some people have the form of DNA with the mutation and others do not, this difference in DNA is called a polymorphism (meaning "multiple forms"—in this context, two or more). And because such *DNA polymorphisms are transmitted from parent to child in the very same way as the alleles that control eye color* (which are, of course, also polymorphisms), they too can be used as genetic markers. Since here the polymorphisms are reflected in different lengths of restriction fragments, they are called restriction fragment length polymorphisms (abbreviated RFLPs and generally referred to as "riflips"). And just as alternative forms of genes are called alleles, so too are alternative forms of riflips or other DNA markers called alleles. In order to make such polymorphisms useful as markers we have to be able to detect them. For an eye-color polymorphism, this is easy: using instruments we are born with, we look at the eyes. Detection of a particular DNA polymorphism—a riflip—also turns out to be easy: using electrophoresis to separate fragments on the basis of their size and radioactively tagged molecular detectors called probes, we look at the DNA.

ELABORATION OF KAN and DOZY's idea that restriction enzyme sites themselves could be used as markers in a genetic map was not long in coming. In a now classic paper published in 1980—"Construction of a Genetic Linkage Map in Man Using Restriction Fragment Length Polymorphisms"—David Botstein, Raymond White, Mark Skolnick, and Ronald Davis generalized and extended this notion, and made what seemed a very bold proposal: to construct a human genetic linkage map based on DNA sequence polymorphisms. Just as Sturtevant had made a map of fly genes based on polymorphisms that could be seen by looking at flies with a magnifying glass, this map of human genes would be based on polymorphisms that could be detected by the partnership of appropriately selected restriction enzymes and probes, combined with a simple instrument for separating DNA fragments by electrophoresis through a gel. It was a start of what is now known as the Human Genome Project.

The initial goal seemed wildly ambitious: to find about ten evenly spaced DNA marker loci (riflips) at intervals along every human chromosome, each acting as a milestone on this chromosome's road of DNA. But the proposal that seemed so wild in 1980 has proved to be stodgy conservatism because of the discovery, within just a few years, of another class of DNA polymorphisms that are much easier to work with than riflips. Called short tandem repeat polymorphisms, or STRPs (pronounced "stirps"), they are actually multiple forms of a repeating pair of DNA bases—C and A—that, for no known reason, are found in tens of thousands of specific places in human DNA. One form might be CACACA, another CACACACA, and yet another CACACACACACACACA; for a given locus there may be as many as ten or more alternatives (that is, alleles). Like other polymorphisms, stirps are inherited, and each can be used as a marker for a particular location on a chromosome. They can all be conveniently detected with a single enzymatic procedure called the polymerase chain reaction (PCR) by using artificial bits of DNA called primers. Because of this technical advance and the wide distribution of stirps throughout the human genome, literally *thousands* of stirp markers have already been identified and mapped to specific chromosomal loci, bringing the intervals between them closer and closer together. Presently an even more convenient technology is being developed to generate a map based on yet another class of DNA polymorphisms called single nucleotide polymorphisms, or SNPs, pronounced "snips," which may soon replace maps based on stirps.

How then can the detailed human genetic map that has been developed be used to find a mood gene? To see how this might work, let us return to our hypothetical family with manic-depressive illness, but drop eye color as a marker. Instead, let us assume that on every human chromosome we have a map of ten stirp markers (*a* through *j* with each marker existing in just two forms, *form 1* and *form 2*), and that the mood-gene locus (*M*) resides on chromosome 1 very close to marker *e*. Let us also assume that the great-grandmother with manic-depressive illness at the top of the hypothetical pedigree on page 102, whose chromosome 1 has *M* (which accounts for her mood disorder) also has *form 2* of marker *e* (that is, *e2*), so it looks like this:

a1	b1	c2	d1	e2M	f1	g1	h2	i2	j1

Since *M* (whose location we don't yet know and are trying to find) is very close to marker *e,* and since great-grandmother has allele *e2* next to *M,* those members of her family who inherited manic-depressive illness from her (via *M*) would all inherit *e2* as well, just as they all inherited the brown eye-color allele in our earlier hypothetical example. By finding that *e2* is consistently inherited along with manic-depression, and that this is not just a chance association (which we would know because of an impressive lod score), we could conclude that the mood-gene locus is very close to marker *e*—in other words, *that they are linked*. And since we would already know the specific address of *e* we would have found the approximate address of *M.*

This, then, is the linkage approach to finding the loci of genes, including mood genes—a critical prelude to the identification and structural analysis of specific genes and the determination of their biological functions. Let me hasten to repeat, however, that in contrast with the hypothetical example in which *M* is a Mendelian dominant allele, the weight of the evidence indicates that mood genes are actually susceptibility genes, which are more difficult to detect by linkage. Furthermore, establishing the approximate location of a locus, as in the hypothetical example, is only the beginning. The intervals between markers in the genetic maps presently used to find such loci are enormous, often millions of base pairs. And though there might well be other markers available between *e* and *f* on chromosome 1 that might help narrow *M*'s location, establishing solid linkage only means that you have identified the suspect's zip code. Finding the street address, and the actual mutation, requires a great deal more skillful detective work.

NEVERTHELESS, DESPITE THESE qualifications, what makes the linkage approach so exciting is that it provides a systematic route to true discovery. Unlike the candidate-gene approach, which is based on hunches about proteins and genes that have already been discovered, the linkage approach is unencumbered by preconceptions. Instead, it can demonstrate a relationship between a disease (such as manic-depression) and a gene whose very existence was not previously known, or a known gene whose involvement in the disease might not have been suspected. Then, as such genes are implicated in the disease process, they themselves becomes tools for the hypothesis-driven science that teaches us which biological or psychological processes they influence. And, as we shall now see, the linkage approach really works.

6

HUNTING
WITHOUT A MAP

In research the front line is almost always in a fog.
—Francis Crick (1988)

When the idea of using structural variations in human DNA (such as riflips) as markers was formally proposed in 1980, the immediate goal was to make a map of every human chromosome to guide the hunt for genes related to inherited diseases. But some scientists became impatient with this methodical approach, which might delay medically relevant discoveries for many years. An alternative was to try at once to establish linkage between a disease locus and whatever markers were already available, even though they covered only limited regions of a few chromosomes. The task would be like hunting for a suspect somewhere in the United States with fragmentary maps of just a few cities. Should the hunters get lucky, the suspect would happen to be hiding in one of those known places. Were he anywhere else, he would elude them.

Among the most ardent advocates of going forward without delay was David Housman, then a young assistant professor of biology at MIT. He had learned about the riflip approach from conversations with one of its inventors, David Botstein, whose laboratory was next door to his. Botstein's main interest was the genetics of yeast, and he was all too aware of the much greater difficulty of doing genetic studies in humans. Housman, however, was eager to give it a try. Even though Botstein urged him to wait for the construction of a map before hunting, Housman decided to take the risk of working with whatever markers were available. He had, in fact, already decided on an inherited human disease that was such a worthy target for investigation that the risk seemed justified.

Also prepared for a risky project were families of patients with that particularly grim hereditary affliction, Huntington's disease, who had banded together in the 1970s to lobby for research on this disorder. Among them was Marjorie Guthrie, whose folksinger husband, Woody, had died of Huntington's disease and whose son, Arlo, was at risk. She was later joined by Milton Wexler, a psychoanalyst trained at the Menninger Clinic, whose wife, Leonore, was also affected, and whose young daughters, Alice and Nancy, each had a fifty–fifty chance of having inherited the abnormal allele. It was Milton who turned this organization, now named the Hereditary Disease Foundation, from an initial interest in seeking new treatments for Huntington's to the hunt for its hereditary basis; and Nancy, a psychologist, who was to play a pivotal role in the hunt itself.

When I first met Milton and Nancy Wexler in January 1978 (early in the year in which I met Michael), the chance of identifying the exact genetic basis of Huntington's disease seemed extremely remote. At a small workshop that the Wexlers organized in Marina del Rey, California, several other consultants and I were asked to help evaluate the progress of the research the foundation was then supporting. A main topic at the meeting was an ongoing project designed to try to find differences between cultures of fibroblast cells from normal people and those from people with Huntington's disease. It was a perfectly good idea to look for such differences, since they could eventually lead to the causative defect (just as the observation that red blood cells sickled eventually led to the defect in hemoglobin). But in the absence of good ideas about what to look for in the cultured cells, the effort was not very likely to succeed. At the dinner party that closed the meeting (given by Jennifer Jones

Simon and Norton Simon, long-time supporters of the Wexlers, at their art-filled Malibu beach home), I had a conversation with a cell-culture expert, Gordon Sato, in which we shared our skepticism about the approach that was being taken: the Wexlers were looking for a needle in a haystack without a powerful technology to aid them in their search.

At the time, of course, neither Sato nor I had heard of riflips. Nor were we aware of the existence of an extraordinary collection of relatives with Huntington's disease who would prove to be perfect subjects for the riflip approach. For despite the rarity of Huntington's disease in the general population (it affects about one person in twenty thousand), there were hundreds of people afflicted with the disorder to be found among the members of a group of large extended families living in northern Venezuela. Clustered in small villages around Lake Maracaibo, a saltwater gulf that connects with the sea, the medical significance of these families had been recognized by a young Venezuelan physician, Americo Negrette, who published a book describing his findings in 1955. As on Long Island, where Huntington's disease was first described, all the affected people in this region of Venezuela were believed to be descended from a single ancestor; in this case, the founder was then thought to be a European sailor who had visited the area about two hundred years ago.

Though Nancy Wexler had known about these families for years, she didn't appreciate their research value until she met David Housman in 1979. To Housman they represented an amazing opportunity to apply the riflip strategy. For one thing, the many dozens of relatives with Huntington's disease should be more than adequate for the statistical studies that would be needed to evaluate attempts to find linkage. Furthermore, because Huntington's disease is so easy to diagnose, and because it is a textbook case of Mendelian dominant inheritance, some of the difficulties that would be expected in genetic studies of more complicated diseases (such as manic-depressive illness) would not arise. Even though construction of a human genetic map was only getting started, all the other conditions seemed so favorable that Housman decided that it was worth trying to find linkage to the Huntington's disease locus with whatever markers he could muster. And Nancy Wexler was more than willing to return to Venezuela with a team of researchers to enlist the cooperation of the afflicted families and bring back blood samples for DNA studies.

To set up a laboratory to examine the blood samples, and to look for new markers, Housman turned to his former doctoral student, James Gusella, who already had some experience in working with DNA. He also helped Gusella to obtain a position in the Huntington's disease program at Massachusetts General Hospital, just across the Charles River from MIT. There, without a map, Gusella started hunting for a riflip that was linked to Huntington's disease.

While the Venezuelan material was being collected in 1982, Gusella did some trial runs with DNA from a family in the United States with Huntington's disease, using the few probes (the radioactive molecular detectors for examining riflip markers) he had by then been able to obtain. This family, which had been studied by Michael Conneally at Indiana University, had too few affected members to lead to a definitive result. But it could give a preliminary indication of the usefulness of the probes Gusella had assembled. To his great surprise, the third probe Gusella tried, which was called *G8*, seemed to have picked up linkage to Huntington's disease: one particular form of the riflip detected by *G8*, which I will call form 1, was more likely to be found in the DNA samples from those relatives with Huntington's disease than in DNA samples from the relatives who were free of the disease. That raised the following question: what were the odds that the co-inheritance of form 1 of *G8* and the Huntington's disease locus was a true indication of linkage rather than a chance association of no real significance?

To answer this question Gusella and Conneally used the statistical method called a lod score (discussed in Chapter 5) and found a value of 1.8. Considering the small number of people with Huntington's disease who were tested, the investigators took this as an encouraging hint—but no more than a hint—

that this riflip might well be near the Huntington's disease locus. The minimum score worth getting excited about is approximately 3, which is generally taken to mean that the odds are about 20 to 1 that the linkage is real. Higher scores, are of course, more convincing; and because the scale is logarithmic (like the Richter scale for earthquakes), an increase of just one unit increases the odds tenfold.

Both Gusella and Conneally recognized that it would take incredible luck to chance upon a marker for the right chromosomal location after only three tries, and they did not allow the lod score of 1.8 to raise their hopes too high. Fortunately, the critical tests could soon be made on the DNA samples from the large Venezuelan families. When those became available, there was no longer any doubt that a Huntington's disease locus existed in the vicinity of the DNA region marked by *G8*. With the Venezuelan samples alone the lod score was a stunningly high 6.7. When the results from the U.S. pedigree were added, the score rose to 8.5, signifying overwhelming odds that the linkage was real. The combination of a wonderful target and incredible luck had led to the validation of a powerful new approach to hunting down the basis of genetic diseases.

Once *G8* was shown to be linked to the Huntington's disease locus, other methods were used to localize the marker (and therefore the disease locus) to human chromosome 4. In their paper in the November 17, 1983, issue of *Nature* that described these results, Gusella and his coworkers concluded that:

> . . . this study demonstrates the power of using linkage to
> DNA polymorphisms to approach genetic diseases for
> which other avenues of investigation have proved unsuc-
> cessful. It is likely that Huntington's disease is only the first
> of many hereditary . . . diseases for which a DNA marker
> will provide the initial indication of the chromosomal
> location of the gene defect.

What the researchers meant by "initial indication" was that even though the DNA sequence that matched *G8* was close enough to the Huntington's disease locus to establish linkage, it still could be far enough—perhaps millions of base pairs—away for the Huntington's disease gene itself to remain elusive. It would, in fact, take another ten years of intensive work by a large team of scientists, the Huntington's Disease Collaborative Research Group, to sift through the upper tip of chromosome 4 and identify this gene (for a hitherto undiscovered and still arcane protein named huntingtin), and the unusual disease-causing variation encoded by the abnormal allele (an expanded region of repeating glutamines in the patients' huntingtin). Nevertheless, by 1983 enough had been discovered to develop a predictive test based on riflips that relatives of a Huntington's disease patient—people like Alice and Nancy Wexler—could take if they wanted to know whether or not they had inherited the genetic abnormality. And because detection of the Huntington's disease gene came so much more easily than expected, it stimulated worldwide interest in searching for other genes that caused human diseases by this method.

NO ONE WAS more excited by these discoveries about Huntington's disease than David Housman. Although he had been confident that the riflip strategy would eventually lead to success, the quick results prepared him to consider working on an even harder problem. Deluged with proposals to study a variety of diseases, Housman became intrigued by the prospect of working on a long shot—the genetics of manic-depressive illness—because it held out the promise of a thrilling payoff: a discovery about the way that human emotions are shaped. And even though it was clear that manic-depressive illness was not a simple Mendelian disorder such as Huntington's disease, it still remained possible that a single locus played a major role in the inheritance of vulnerability to this disease. Were this the case, the riflip approach should be able to find it.

What persuaded Housman to go forward with this project was the availability of a large and isolated community, the Old Order Amish of Lancaster County, Pennsylvania, in which familial patterns of manic-depressive illness had already been examined. Further study of this group might do for this mood disorder what the families that lived around Lake Maracaibo had done for Huntington's disease. Descended from a few hundred Swiss and German Mennonite founders who came to America to seek religious freedom in the eighteenth century, their high birthrate had raised their numbers to about twelve thousand. Although the Old Order Amish as a whole did not have an unusually high incidence of manic-depression, many of those affected were concentrated in particular families, some of whom had been identified over the prior two decades by a team of scientists led by Janice Egeland.

Egeland had learned about the great potential value of the Amish for genetic studies while working with Victor McKusick, who had founded a pioneering program in medical genetics at Johns Hopkins University in 1957. Trained as a cardiologist, McKusick was drawn into medical genetics because of an interest in the cause of the heart abnormalities in Marfan's syndrome, the hereditary disorder that also produces long limbs and fingers and that may well have affected Abraham Lincoln (McKusick believes the chances of that are about fifty-fifty). Medical genetics was such a tiny and obscure field in the 1950s that McKusick had to band together with a few colleagues to found the Galton-Garrod Society to promote its development. His greatest contribution has been a compendium, *Mendelian Inheritance in Man*, first published in 1956 and continuously updated, which played a major role in the development of a human genetic map even before the discovery of markers based on DNA polymorphisms.

One reason McKusick became interested in the Amish is that he knew that rare inherited diseases found in a community descended from a small number of ancestors (and without marriage to outsiders) are often derived from a single individual—the founder. In such cases genetic studies may be facilitated in several ways, the most obvious of which is that the disease-causing mutation would be identical in all the affected individuals (in contrast with a disease in more diverse populations, in which the same disease might arise because of mutations in any one of several different genes—that is, genetic heterogeneity). In fact, Egeland's early work with Mc-Kusick included studies of inherited forms of dwarfism that are probably attributable to founders in the Pennsylvania Amish community.

In the course of visiting and interviewing relatives in the Amish farmhouses during the 1960s, Egeland became fascinated with their old-fashioned and deeply religious way of life. In sticking with the black clothes and horses-and-buggies of their ancestors, they had kept themselves apart from the cultural upheavals that surrounded them. And even though as a geneticist Egeland was a practitioner of a very modern activity, many Amish people grew comfortable with her, divulging family secrets and even sharing stories about the wayward conduct of their neighbors. In this way Egeland learned that the Amish had noticed that forms of misconduct—from buying "fancy" clothes to a tendency to suicide—often ran in families. In fact, about half of the twenty-six suicides among the Amish of south-eastern Pennsylvania between 1880 and 1980 occurred in only two families. When Egeland proposed to study the genealogy of mood disorders, many Amish families were eager to participate. Despite their abhorrence of the frivolous gadgetry spawned by new technology, genetic studies offered something potentially useful: a way to detect individual vulnerabilities that raised the possibility (and, therefore, the responsibility) of taking preventative measures.

By the mid-1970s Egeland's informal studies had progressed to the point that she had obtained support from the National Institute of Mental Health for a systematic study of mood disorders in the Amish. Her main aim was to study families with members who had obvious signs of manic-depressive illness, seeking to find out which other relatives might also be affected. Imagine her excitement when, just as she had collected several large pedigrees with a high concentration of mood disorders, the linkage of *G8* to Huntington's disease was reported. Knowing of Housman's role in this work, Egeland hoped to persuade him to help her to apply the same approach to manic-depressive illness.

Housman was fascinated by Egeland's proposal because her pedigrees provided such strong evidence that genes play an important role in manic-depression. But he was also wary, because the pattern of transmission of this mood disorder was a great deal murkier than that of Huntington's disease. In contrast to Huntington's disease, which is transmitted by a single mutant allele, manic-depressive illness is a complex trait that may involve the interaction of alleles of several mood genes as well as a variety of environmental factors. Though an optimist by nature, Housman was well aware that finding a locus that influenced vulnerability to manic-depression—that is, a susceptibility gene—might be out of reach of the available technology. If Huntington's disease had taken luck, manic-depression might take a miracle.

Nevertheless, given the material that Egeland could provide, Housman decided that applying the riflip approach was worth a try. So did Daniela Gerhard, a postdoctoral student who had joined his laboratory initially to study the role of globin genes in the commonest form of hereditary anemia, thalassemia ("anemia of the sea," so called because of its relative frequency among Mediterranean peoples). Because she and Housman were already engaged in mapping probes on chromosome 11, where the beta-globin genes are located, they decided to use the same probes to look for linkage to the manic-depression observed in the Amish. They were, of course, aware that using only these probes would greatly limit their chances of finding linkage, because chromosome 11 contains only a small fraction of the total human genome; but it was the most convenient place for them to start.

I FIRST LEARNED of this project in 1983 when, stimulated by my talks with Michael and news of the success with Huntington's disease, I organized a small conference—"Looking for

Genes Related to Mental Illness"—at the Neurosciences Institute, then housed at Rockefeller University. Because Huntington's disease would serve as a model for this new approach to psychiatric genetics, one of the first people I called was Nancy Wexler. From her came the news that, far from being a pipe dream, application of the riflip approach to mood disorders was already under way in the laboratories of Housman, Gerhard, Egeland, and other collaborators, so that their participation would bring real hands-on experience to our discussions.

By the time the conference was convened in October 1984, Gerhard and colleagues, much to their own surprise, had obtained tantalizing results that suggested that a gene involved in manic-depression might actually be located in just the place they had been looking—chromosome 11. Although the data were presented with great caution, everyone was excited by the possibility that riflips could be so easily applied to what we all had expected to be an exceptionally difficult problem. But DNA samples from many more Amish patients would have to be examined to follow up this lead, and the researchers decided that this could be best accomplished by focusing on their most promising pedigree, known as pedigree 110.

What distinguished the family depicted by pedigree 110 was its very high concentration of manic-depressive illness. On careful examination by a formal diagnostic procedure, 19 of 81 members had been found to have some form of this broadly defined mood disorder. Of the 19 affected people, 14 had episodes of both mania and depression and 5 had severe depression. Assuming (with Kraepelin) that these were alternative manifestations of the same hereditary disease, its expression in almost one in four members of this pedigree raised the possibility that they shared a dominant allele of a mood gene that was about 50 percent penetrant (so that about half of those who inherited this allele actually showed signs of manic-depression—broadly defined—whereas the other half, who would have inherited this same allele, seemed unaffected). Viewed in this way, pedigree 110 was an acceptable target for linkage analysis.

Two years later the DNA results were in, and they immediately attracted worldwide attention. Two probes, named *HRAS1* and *INS*, both of which served as markers for the same general area of chromosome 11, now gave fairly convincing evidence of linkage to a locus for manic-depressive illness, opening the path to the identification of a mood gene.

With *HRAS1* alone there was an impressive lod score of 4.1. Although considerably lower than the score found with *G8* in Huntington's disease, it comfortably exceeded 3.0, the accepted statistical criterion for linkage. A lod score of 4.1 (which indicates odds of about 200 to 1 that the linkage is not due to chance) was, in fact, surprisingly robust, given the possibility of diagnostic errors and the evidence of incomplete penetrance, both of which would diminish the chances of finding linkage. Furthermore, the finding with *HRAS1* was supported by the finding that *INS*, a nearby marker, was also linked to manic-depression, with a lod score of 2.6. The lower lod score with *INS* could simply mean that the site it marked was farther from the mood gene than the site marked by *HRAS1*.

Despite this convergent evidence, there was reason to be worried. As Gerhard and colleagues knew from the start, there was only a small chance of finding linkage to the limited region of the genome that they had arbitrarily chosen to examine. Although they had also tested bits of a few other chromosomes, with negative results, the vast remaining territory of the genome had been left unexplored. Had it not been for the earlier stroke of luck with Huntington's disease, everyone might have been more skeptical. But since Gusella had been so lucky, why not Gerhard as well?

The combination of this line of reasoning and the persuasive results accounted for the great enthusiasm that I and others expressed to Housman when asked to comment on a draft of the manuscript that was being prepared for publication. Although everyone understood that the results could only be taken to mean that there was a good chance that a mood gene was located in chromosome 11, and that proof would require the subsequent identification of the allele associated with manic-depression, it was difficult to quarrel with the bold summary of the paper by Egeland, Housman, Gerhard, and their colleagues that was published in *Nature* on February 26, 1987: "An analysis of the segregation of restriction fragment length polymorphisms in an Old Order Amish pedigree has made it possible to localize a dominant gene conferring a strong predisposition to manic depressive disease to the tip of the short arm of chromosome 11."

BUT THERE WERE already signs of trouble, in the very same issue of *Nature*. Along with the evidence of linkage in the Amish, the journal published two reports by other groups that, knowing of the preliminary findings that Gerhard had made public, had both tried and failed to find linkage of manic-depressive illness to the same region of chromosome 11 in other pedigrees. Even worse, their linkage analyses, which used exactly the same probes, *HRAS1* and *INS,* produced lod scores that were sufficiently negative actually to exclude the region of chromosome 11 that Egeland and her colleagues had linked to manic-depression.

Why then did *Nature*, one of the most highly regarded science journals, simultaneously publish both a major article announcing a dramatic positive finding and two short articles that appeared to refute it? The answer was provided in the same issue by a *Nature* editor, Miranda Robertson, in her commentary on this body of work. Following the lead of the authors of both negative papers, Robertson concluded that the results must indicate that manic-depression is "genetically heterogeneous." By this she meant that a vulnerability to the development of this mood disorder was apparently controlled by alleles of either of two (or more) entirely different genes, one located in the region identified by Gerhard on chromosome 11 and at least one other located elsewhere; and that inheritance of the critical allele of any one of these mood genes might play a central role in vulnerability to manic-depressive illness. In support of her view that there are multiple genetic sources of manic-depression, she called attention to another paper that would appear in *Nature* two weeks later, this one providing evidence for a different mood gene that had been localized to a region of the X chromosome.

No one was really surprised by the proposal that one of a number of genes might play a decisive role in particular families with manic-depression. If phenotypes such as wrinkled peas can be traced to alleles of different genes, why not a serious mood disorder? For this reason there was general support for Robertson's view that *Nature* was publishing not just one exciting result, the positive linkage in the Amish, but also a second—evidence for the genetic heterogeneity of manic-depressive illness.

Plausible though this interpretation was, it was soon shattered by new data from Amish pedigree 110. Published in *Nature* on November 11, 1989, these results were obtained by the original group of investigators, including Egeland, Housman, and Gerhard, now joined by new collaborators from the National Institute of Mental Health.

The first problem arose when, on periodic reexamination of the members of this pedigree, two previously asymptomatic people had developed serious mood disorders, one a manic episode and the other a major depression. When the lod scores for *HRAS1* and *INS* were recalculated in the light of these changes, both plummeted to well below 2, a value that did not meet the criterion for linkage. Making matters worse, examination of other branches of pedigree 110, which also had a high concentration of manic-depressive illness that presumably reflected the same mood-gene allele, gave strongly negative results for linkage to chromosome 11. Simply adding results from one six-member branch, consisting of two people with bipolar disorder and four who were unaffected, brought the lod scores down close to zero. When data from a 31-member branch with 4 people with bipolar disorder and 4 with major depression was also added, the lod score descended to such negative depths that linkage of *HRAS1* and *INS* to manic-depressive illness in the newly enlarged pedigree 110 could actually be excluded.

How could the original conclusion, which had so comfortably met the accepted criteria for linkage, prove so fragile? The widely held opinion was that, instead of having had the very good luck to have picked a marker—*HRAS1*—that happened to be very close to a mood-gene locus, the original group of investigators actually had had the very bad luck to have been led to an erroneous conclusion by an inherent property of the lod score method. Though the odds against coming up with a lod score of 4.1 by chance are about 200 to 1, this study was apparently that 1 case in 200. As Miranda Robertson put it in commenting on this new article, "the simplest explanation is that the apparent linkage to a chromosome 11 locus was in fact just chance."

DISAPPOINTMENT HAS ALSO replaced the claim for linkage of manic-depressive illness to the X chromosome, which Robertson had mentioned in her first commentary on the Amish studies. That claim, published in *Nature* on March 19, 1987, by a team led by Miron Baron of Columbia University, was based on seemingly compelling evidence that the mood disorder was linked to two markers that had both been mapped on the X chromosome by the pre-DNA techniques of Morgan and Sturtevant. But in this case too the evidence for linkage could not withstand subsequent reexamination.

This team had become interested in the possible role of the X chromosome in manic-depression because they had found several large Israeli pedigrees with both a high concentration of this mood disorder and another distinctive property: there were no cases in which both a father and his son were affected. As T. H. Morgan could tell you, this would be expected if the mood disorder were caused by a dominant allele on the X chromosome, because fathers never transmit their X chromosome to their sons (who always get their father's Y). So fathers with manic-depression could transmit the disorder only to their daughters, whereas affected mothers could transmit it to children of either sex, since each receives a maternal X chromosome.

There was, however, good reason from the start to be cautious about X-linked transmission of manic-depressive illness: it is very unusual to find large pedigrees with this disorder that do not show some evidence of male-to-male transmission. But the attractive aspect of the X-linkage proposal was that, at a time when a riflip map of other chromosomes had not yet been constructed, the X chromosome was already familiar territory because the chromosomal loci of many X-linked diseases had already been mapped using Sturtevant's classic procedure. Among the genes on this chromosome, one encodes a protein involved in color vision with mutant alleles that result in color blindness; and another, about 2 centimorgans away, encodes an enzyme, G6PD, with mutant alleles that may give rise to a disease called favism that reflects a deficiency of this enzyme in red blood cells. Should alleles for color blindness or G6PD deficiency be regularly inherited along with manic-depressive illness, this would prove the existence of an X-linked form of manic-depression and even point to the location of the mood gene.

The pedigrees for Baron's X chromosome study were all derived from the patient population of the Jerusalem Mental Health Center. Among the Israelis served by this facility are members of several large families, mostly of non-Ashkenazi (that is, Middle Eastern) Jewish origin, with a high concentration of mood disorders. Included in these families were patients with many different types of mood disorders, ranging from full-blown manic-depressive illness—that is, bipolar disorder, type I—to cyclothymic disorder, a milder condition characterized by mood swings that do not produce significant disability. All these conditions were assumed, for the purpose of the study, to be reflections of the same mood-gene allele. In addition to its large collection of potential subjects for genetic research, the Jerusalem Mental Health Center also has access to data on color blindness and G6PD deficiency, which the Israeli army routinely tests for in new recruits. By choosing families known to have both a mood disorder and either color blindness or G6PD deficiency, the researchers could look for linkage between the locus that controls vulnerability to the mood disorder and the region of the X chromosome covered by these adjacent markers.

The initial results were striking. In the non-Ashkenazi pedigrees the broadly defined mood disorder was apparently linked to either color blindness or G6PD, with startling lod scores—as high as 7 to 9 if computed after making certain favorable assumptions. Since the gene for G6PD and the locus for color blindness are so close together on the X chromosome, this justified the combination of the results with these two markers as representative of a clearly defined region of the X chromosome. Somewhere nearby, there was presumed to be a mood gene.

But there were also some significant problems that diminished the impact of this strikingly positive result. For one thing, the single Ashkenazi (European Jewish) pedigree that the team examined showed no evidence for linkage to this region of the X chromosome. Combining the Middle Eastern Jewish pedigrees and the Ashkenazi pedigree—which had also been selected for study because of absence of evidence of male-to-male transmission—substantially lowered the lod scores, necessitating the plausible explanation of genetic heterogeneity. More disturbing was the finding that if two individuals with cyclothymic disorder in the Middle Eastern Jewish pedigrees were classified as being unaffected rather than affected (by simply setting a more stringent diagnostic criterion), the lod scores plummeted to 4.4 for linkage to color blindness and to an even lower level—2.0—for linkage to G6PD.

More bad news was to come when the initial study, which had relied on clinical tests of color blindness and measurements of the enzyme activity of G6PD, was redone with the DNA markers that became available in the early 1990s. In a follow-up study using these DNA markers, the results of which were published by Baron and a group of new collaborators in the January 1993 issue of *Nature Genetics,* attention was confined to only three of the large Middle Eastern Jewish pedigrees that had each shown strong evidence for linkage. When taken together the results from these three pedigrees no longer supported linkage to the X chromosome. And no explanation could now be offered for the absence of father-to-son transmission that had inspired this whole study.

THESE FALSE STARTS may seem to cast doubt on the applicability of the linkage approach to complex disorders such as manic-depressive illness. But limitations of both the Amish and Israeli studies did not really allow for a fair test of this approach. One such limitation was that the criteria for classifying a person as manic-depressive were very broad. Included were not only people who had had episodes of full-blown mania as well as depression—that is, people with unequivocal evidence of having bipolar disorder, type I—but also people who had only depression or (in the Israeli study) a milder condition, cyclothymia. Whereas some people with these other diagnoses may well share relevant alleles with those affected with the bipolar disorder, others may not. Had only the people affected with the most severe condition (bipolar disorder, type I) been considered in the initial studies, neither research group would have found significant positive results.

Another reason that the Amish and Israeli studies were not a fair test of the linkage approach is that they were done without a chromosome map. As a result, only parts of one or a few chromosomes were tested for linkage, whereas the relevant gene (or genes) could be anywhere. Since 1987, when these initial studies were published, important technical advances have greatly facilitated work of this type. For one thing the relative ease of measuring the DNA variations called stirps has provided a detailed map of all human chromosomes, replete with thousands of closely spaced markers, which offers the possibility of quickly finding the rough location of a relevant gene instead of just hoping to stumble upon it. Furthermore, alternative experimental and statistical methods have been developed to supplement or even replace the lod score approach, which only works well for typical Mendelian traits.

In view of these developments there have been striking changes since the late 1980s in the tactics of hunting for mood genes. The most obvious is that the hunt now begins by deploying a large enough battery of markers to screen all human chromosomes. In practice this presently means that the DNA of every individual in the study is screened with about three hundred to five hundred stirp markers in order to try to find the approximate address of the locus of interest. As a map based on snip markers is developed, the hunt will be made even easier. It has also become clear that it is essential to use stringent criteria in defining affected individuals, since even a single error in diagnosis can defeat an attempt at establishing linkage. This is particularly important in the case of manic-depressive illness, which, in the broad Kraepelinian view, shades into normal mood variations. For this reason it is safest to start new hunts for mood genes by considering only those people who have the form of manic-depressive illness that is most clearly both familial and distinctive—bipolar disorder, type I.

But a map full of markers and stringent diagnostic criteria is not all we need to hunt for mood genes. To make possible a successful linkage study we also need the cooperation of a sufficient number of affected relatives.

As with Huntington's disease, this requirement would be met in Latin America.

7

ANA'S FAMILY

The work of . . . statistician[s] is that of the Israelites in Egypt. . . . They must not only make bricks but find the materials.

—Francis Galton

The image most Americans have of Costa Rica is of a luxuriantly tropical vacation spot—not the sort of place that would lend itself to systematic psychiatric research. Yet this tiny country has provided us with exceptionally valuable knowledge about an extended family whose pedigree is riddled with manic-depressive illness, opening an important path in the hunt for mood genes. Identification of this family came about through a search for help by a concerned relative I will call Ana.

BORN IN COSTA RICA to wealthy and intellectual parents, Ana was raised in great comfort and received an international education. Tall and fair, she has a dazzling smile, an infectious vivacity, and an obvious intelligence. While traveling in Europe she met and married a shy American engineer, Richard, with whom she settled in a suburb of New York City. In time, some other members of her family also moved to the United States, but most remained in Costa Rica, where Ana visited frequently.

Ana and Richard had two sons, William and John. William, the elder, was a particularly sparkling and happy child, a very good student, well liked by classmates and teachers. In high school he was the goalie on the soccer team, dated extensively, and played the clarinet. Like his father, who had gone on to become a founder of a successful computer software company, William was interested in mathematics. He had begun to think of a career in aerospace.

Then, at sixteen, William became manic. He had just begun his junior year at an academically demanding private school and was staying up most of the night, pacing back and forth in his room. At first Ana and Richard dismissed this as a normal response to a heavy load of classes. But it was harder to overlook his newfound boastfulness and his odd, elated excitement. When Ana tried to talk to him he couldn't sit still and kept switching topics even as he tried to assure her that everything was OK.

What made William's behavior especially disturbing to Ana was that she knew these signs all too well. While she was living at home in Costa Rica, her older brother Pedro, then in his early twenties, had become agitated, belligerent, and extremely moody. Eventually he was hospitalized, given shock treatments, and diagnosed as having manic-depressive illness. Only years later, after several more attacks, was he treated with lithium, which brought some relief. At about that time a much younger brother, Juan, was similarly affected, and received the same diagnosis and treatment. Their mother, Esperanza, had also been hospitalized because of a serious mood disorder with episodes of both depression and full-blown mania. Ever since the psychiatrist in Costa Rica who first treated Pedro with lithium had pointed out to Ana the hereditary nature of manic-depressive illness, she had been living in dread of the possibility that her sons might be affected. Though Ana herself had apparently escaped, it seemed to be William's turn to assume the family burden.

The psychiatrist who interviewed William confirmed Ana's suspicion. After two long meetings and a battery of medical tests he was confident of the diagnosis: bipolar disorder, type I. Fortunately he was able to treat William successfully as an outpatient with medications that calmed him until lithium could take effect, thereby avoiding the need for a disruptive period of hospitalization. With this treatment, combined with frequent sessions of counseling, William was soon back in school and doing well. But despite his rapid recovery, William's diagnosis implied that he would require treatment for the rest of his life.

Once Ana's fears for William had been realized, she accepted his illness with remarkable equanimity. Knowing that her mother and brothers had been able to live productive lives between attacks, Ana felt very satisfied with the treatment that William was receiving and continued to be optimistic about his future. Nevertheless, she knew that William would always be in danger of another attack, and grew increasingly apprehensive that the same fate might befall her other son, John. It was for these reasons that Ana discussed the details of William's illness with her friend Wendy, a social worker employed at a psychiatric clinic at Columbia University's College of Physicians and Surgeons.

Although Wendy had not previously paid much attention to the familial nature of manic-depressive illness, she was startled to learn of the frequency of mood disorders in Ana's family and concerned that Ana might have guilty feelings about possibly having transmitted a hereditary disease to her son. Wendy therefore suggested that Ana visit her clinic to talk over the special issues raised by William's illness and John's vulnerability. It was in this context that, in the fall of 1988, Ana found herself telling the stories of William and Pedro and Juan and Esperanza to a young postdoctoral trainee in the clinic named Nelson Freimer.

I FIRST MET Nelson Freimer in 1986, at the University of California School of Medicine in San Francisco (UCSF), where he was beginning his final year of clinical training as a resident in psychiatry. I had just moved to UCSF to head its department of psychiatry and to build a research program devoted to molecular and genetic approaches to mental illness. Having established itself as a world-class biomedical research institution in the 1970s (aided in no small measure by my mentor, Gordon Tomkins, who played a vital role in its development between

1969 and his untimely death in 1975), UCSF was an ideal place to launch this new program. Concerned that the psychiatry department had not kept up on this front, UCSF's dean, Rudi Schmid, was eager to see it take a major step in this direction and was prepared to invest millions of dollars for new laboratories and new faculty.

But there was still a missing ingredient: very few psychiatrists had been trained to do the type of laboratory-based research that was needed. Although the foundations for an attack on serious mental illnesses were being laid by a broad range of biological scientists, there was a pressing need for a crop of young psychiatrists who would have a foot in each camp, bringing new technologies from biology to bear on psychiatry's complicated clinical problems. So the trick would be to find talented psychiatry residents who were interested in postdoctoral training in biological research as a prelude to joining the faculty. Of the dozens of psychiatry residents at UCSF, Nelson Freimer was clearly the most promising candidate.

Fortunately Nelson was already thinking about seeking additional scientific training. Having decided to work in psychiatric genetics, he had whetted his appetite with a small project on the relationship between familial thyroid disease and mood disorders, with the assistance of his faculty advisor, Victor Reus, and Mary-Claire King, a geneticist then working across the bay at the University of California campus at Berkeley. But the field of human genetics was becoming more and more complicated, and Nelson was growing concerned that his lack of formal laboratory experience would stand in his way. While in medical school at Ohio State University, he had considered and rejected concurrent M.D. and Ph.D. training, which was becoming the established route to a biomedical research career, because he preferred talking with patients to working at the lab bench. Would he be comfortable in the laboratory? Was it too late for him, already twenty-nine years old, to acquire the requisite skills?

My advice to Nelson was that the best way to find out was to spend a few years working in a laboratory, the route I had taken many years before. The same advice came from David

Cox, a young medical geneticist at UCSF to whom I had turned for help in creating a program in psychiatric genetics and in overseeing the construction of a laboratory for genetics research in a large abandoned kitchen of UCSF's Langley Porter Psychiatric Hospital. Our hope was that after he completed his postdoctoral training Nelson would return to UCSF as a junior faculty member, and that he would share this new laboratory with David and a molecular geneticist, Rick Myers, who had also become interested in studying mental disorders.

What clinched Nelson's decision to get some laboratory training was his meeting with Conrad Gilliam, yet another budding young geneticist, on a hot summer day in Boston in mid-1987. Nelson had gone to Boston to interview a patient as part of his project on familial thyroid disease and mood disorders. He had decided to become a postdoctoral trainee at Columbia University and was planning to work with Myrna Weissman, a distinguished psychiatric epidemiologist (whose studies of early-onset depression were mentioned in Chapter 2). Gilliam, then a postdoctoral trainee in James Gusella's laboratory, was working on the hunt for the Huntington's disease gene and had accepted a junior faculty position at Columbia. When he heard this, Nelson arranged to visit Gilliam in Boston.

Their meeting was so successful that Gilliam offered to provide Nelson with the laboratory training he needed when they both moved to Columbia; and Nelson was so captivated by the prospect that he immediately accepted. Nelson recalls that what he found so appealing about this meeting was Gilliam's demeanor—deeply committed to his work but also faultlessly casual, he was dressed only in a white lab coat and jogging shorts, shirtless on that hot summer day. Here was a dashing and enthusiastic laboratory scientist, not many years his senior, whom Nelson was ready to emulate.

When Nelson arrived at Columbia in the fall of 1987, Gilliam had already begun a collaboration with Miron Baron, the scientist in charge of the Israeli project on X-linked manic-depressive illness. Their goal was to use DNA markers to evaluate the linkage that had been inferred from examinations of color blindness and levels of G6PD, in the hope of determining the location of the mood gene more precisely with the methods that had worked for Huntington's disease. Assisting Gilliam and Baron was a perfect opportunity for Nelson to learn how to use DNA probes to examine riflips. If the chromosomal location of a mood gene was clearly established, Nelson would learn the additional techniques that would be needed to narrow down the location even further—from zip code to approximate street address. Then he would participate in the laborious task of determining the exact sequence of DNA base pairs in this "neighborhood," thereby completing the identification of the mood genes.

But instead of the exercise in successful gene-hunting that he had expected, Nelson soon found himself in an extremely frustrating situation: the work with DNA markers in conjunction with Gilliam and Baron failed to support the initial results. Although it would take Nelson a few years to give up on this project, his hopes for it were quickly tempered. Though Gilliam could shrug off the disappointment because there were other problems he was interested in studying, Nelson was severely shaken. Perhaps his decision to go into psychiatric genetics had been a big mistake.

Yet the evidence that manic–depressive illness has a genetic basis remained unchallenged. Nelson soon came to realize that the failure to confirm the initial results indicated not that genetic studies should be abandoned, but that a more careful and systematic approach would be required—both in patient selection and in attempts to establish linkage. Despite his crushing experience, Nelson decided to start again from scratch, finding families of his own to work with.

Enter Ana.

LIKE MANY PEOPLE born in Costa Rica, Ana traces her ancestry to the eighty-six Spanish families who settled there between about 1569 and 1600. The men were mostly soldiers and adventurers who had participated in the conquest of Mexico and had been rewarded with land in nearby territories. The region they occupied is in Costa Rica's Central Valley, which is surrounded by chains of mountains that shield it from the Caribbean to the east and the Pacific to the west. Because this rocky terrain had few parcels of land that were suitable for traditional agriculture, not many families followed the small number of early settlers. As a result, these few hundred Spanish immigrants became the main progenitors of more than a million living Costa Ricans who trace their origins to the Central Valley.

There were native Amerinidians in the Central Valley when the Spanish arrived, and they too are significant progenitors of contemporary Costa Ricans. But although initially they outnumbered the Spanish settlers, their contribution to the current gene pool was restricted by many deaths caused by violence and infectious diseases. According to records kept by the Spanish crown and the Roman Catholic church, the numbers of the Amerindians were not very large to begin with—there were only about 4800 in the Central Valley when the Spanish arrived, and the population had declined to about 1400 in the year 1700. By then there were 2100 Spanish, virtually all of them descended from the founding families because at that time there was little interbreeding between the Spanish and the Amerindians.

Thereafter interbreeding became common, and the population in the Central Valley grew to about 250,000 people by 1900, almost all descended from the small number of Spanish and Amerindians who had been there two centuries before. Although there was substantial immigration into Costa Rica in the late nineteenth century, it consisted mostly of people from the Caribbean, largely of African descent, who settled in the eastern coastal regions where banana plantations were established. These newcomers were explicitly excluded from the Central Valley, which had taken on a distinctive Spanish colonial character.

A great deal is known about the genealogy of many contemporary Costa Rican families from church records and other sources that have been collected and analyzed over the years. In recent times this work has been extended by Eduardo Fournier, a retired professor of history at the University of Costa Rica who remains passionately interested in poring through dusty volumes to reconstruct family trees. The availability of these family trees and the small size of the ancestral pool in the Central Valley in 1700—more than a dozen generations ago—have made Ana's extended family especially valuable for genetic research on manic-depressive illness.

But Nelson knew none of this history when he first met Ana in the clinic at Columbia toward the end of 1988. Nevertheless, impressed by Ana's story, he interviewed a few of her relatives who had moved to the United States, who substantiated her observations about mood disorders in the family. But these were only anecdotal reports. Nelson was well aware that, if Ana's family were to be truly valuable in the hunt for mood genes, these reports would require confirmation by formal interviews of a great many relatives in Costa Rica. Such a large-scale venture would require substantial funding, the assembly of a team of specialists, and research collaborators in Costa Rica who could help track down and interview Ana's relatives in their own language.

To obtain funding to mount a research program on the genetics of manic-depressive illness, David Cox, Rick Myers, and I had in 1989 decided to submit a grant proposal to the National Institute of Mental Health (NIMH) for five years of support. Since David and Rick were mainly interested in developing new techniques for narrowing down the locus of a mood gene, they proposed to do their initial work with DNA samples already available from Egeland's Amish study. But when we learned about Nelson's promising contacts with Ana and her relatives, and his interest in returning to UCSF in 1990 to work on this project with Victor Reus, we expanded the grant proposal to include support for a Costa Rican component. And even though NIMH provided us with funds that were initially earmarked only for the Amish work, arguing that more pilot work would have to be done with Ana's family in order to justify additional financial support, we soon obtained permission to apply a portion of the funds that we had been awarded to the work in Costa Rica.

JUDGED BY ITS potholed and unpaved roads, Costa Rica gives the impression that it is an underdeveloped country. In fact, it has many advanced features, one of them its system of psychiatric care. Centralized in the capital, San José, this national system treats all people with serious mental illness in either the National Psychiatric Hospital or Calderon Guardia Hospital, making the registries of these hospitals an invaluable source for data on familial patterns of manic-depressive illness.

The existence of such a well-organized and effective system of psychiatric services is due in part to a remarkable family of psychiatrists whose present senior member is Alvaro Gallegos. Like his grandfather (who studied psychiatry in Kraepelin's clinic in Munich), Alvaro received some of his medical education in Europe. He followed that training with a residency in psychiatry at Johns Hopkins and has continued close contact with modern developments in psychiatry in the United States.

It was therefore natural that when Nelson began visiting Costa Rica early in 1990, Alvaro Gallegos would be one of the first people he would seek out. Fortunately Alvaro was himself extremely interested in manic-depressive illness, and quite aware of the families in which it was concentrated. So too was Alvaro's protégé, Luis Meza, who was in charge of the lithium clinic that Alvaro had established for the treatment of patients with manic-depression at Calderon Guardia Hospital. The lithium clinic would soon become a major source of subjects for genetic studies.

On this initial trip to Costa Rica, Nelson also had lengthy discussions with another member of this growing team, Pedro Léon, a professor of genetics at the University of Costa Rica. Nelson was greatly impressed with Pedro's work on a form of hereditary deafness in a Costa Rican family because it illustrated the value of studying inherited human diseases in the members of the Costa Rican population who were descendants of the small number of Central Valley founders. Working with

Mary-Claire King at Berkeley, Pedro was well on his way to localizing the gene that is responsible for this rare Mendelian dominant form of deafness by taking advantage of the fact that all those affected with it could be traced back to a common ancestor. This ancestor, Felix Monge, who was born in the Central Valley in 1754, was himself a known descendant of a family that had immigrated to Costa Rica from Jerez de la Frontera, Spain, about 1600. Because of Pedro's experience in conducting a family study and his contacts in the Costa Rican academic community, he would play an important role in establishing Nelson's project. He also agreed to provide laboratory facilities at the University of Costa Rica for initial processing of the DNA samples, and to help with the DNA analysis.

Another critical collaborator whom Nelson met on this trip was Pedro's wife Mitzi Spesny, a clinical psychologist. Although in the ninth month of pregnancy during Nelson's visit, Mitzi was so intrigued by the prospect of a study of the genetics of mood disorders that she volunteered to take part in the initial meetings with Ana's family, driving long distances on dirt roads in her four-wheel-drive automobile to track some of them down. Mitzi is now the clinical coordinator of the Costa Rican component of this research program.

WHILE PATIENTS WERE being approached in Costa Rica, other clinical components of the project were being organized by Nelson's former faculty advisor, Victor Reus. He was soon joined by Michael Escamilla, a psychiatry resident at UCSF who had become interested in psychiatric genetics while still in medical school at the University of Texas in Dallas. Of Mexican-American parentage, Michael's fluency in Spanish was enormously valuable: he could conduct personal interviews with patients and their relatives in Costa Rica directly, without having to depend on an interpreter.

This fluency would also be put to use in Michael's preparation (with Victor Reus, Mitzi Spesny, and Lilia Solorzano, a Costa Rican psychiatrist) of a Spanish version of a standard interviewing tool, the Schedule for Affective Disorders and Schizophrenia (SADS). The Spanish-language SADS was then used by Mitzi and other experts to evaluate patients in Costa Rica. Those who received a diagnosis of a major mood disorder based on the SADS were interviewed a second time, by a different expert, usually Michael Escamilla, using the translated version of another tool, the NIMH Diagnostic Instrument for Genetic Studies (DIGS).

To assign a formal diagnosis the data from both the SADS and DIGS assessments were then independently evaluated, along with the patient's detailed psychiatric records, by two psychiatrists at UCSF, using a formal procedure that gave a result called a best-estimate diagnosis. Once a diagnosis was established (by agreement of the independent evaluations of the two "best estimators"), the patient's DNA was examined. Testing of markers for most of the chromosomes was coordinated by a UCSF psychiatry resident, Alison McInnes. The remaining chromosomes were screened in Costa Rica by a team of scientists led by Pedro Léon.

Evaluation of the data from the diagnostic studies and their relationship to the DNA results required yet another type of specialist, one skilled in statistical techniques—the project's Francis Galton. For this role Nelson recruited a young Dutch geneticist, Lodewijk Sandkuijl, whom he had met while both were doing postdoctoral work at Columbia. From his offices in three Dutch universities, Lodewijk supervised the statistical aspects of the project mainly by electronic transmission of data and analyses.

By the middle of 1992 the clinical results from Ana's family were in. As expected, this formal procedure confirmed the impression that she had many relatives with a major mood disorder, as displayed in the pedigree that is reproduced on the following page.

Only a fraction of the total number of Ana's affected and unaffected relatives is shown, to help protect the anonymity of the family. Those marked as affected were all evaluated by the formal diagnostic procedure described, with the exception of

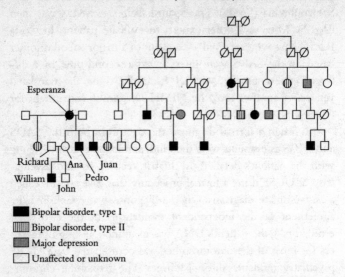

Bipolar disorder, type I
Bipolar disorder, type II
Major depression
Unaffected or unknown

two deceased family members (their symbols marked with a slash) who were diagnosed on the basis of a lifetime of illness that was extensively documented in their hospital records. One of these deceased relatives is Ana's mother, Esperanza.

Of the members of Ana's family with serious mood disorders, fifteen who are still alive are shown on the pedigree: eight with bipolar disorder, type I (episodes of full-blown mania and of major depression), four with bipolar disorder, type II (episodes of mild mania and of major depression), and three with major depression (without any mania). But because full-blown manic episodes, the hallmark of bipolar disorder, type I, are generally accepted to be the most clearly hereditary of all forms of mood disorder, only those with this diagnosis were included in the linkage studies. This decision was made from the start, even though it seemed likely that most of the relatives with the other forms of mood disorder would share some or all of the same mood-gene alleles—a matter that will be revisited when the alleles at work in this family are identified.

Even using this very restrictive criterion, Ana's family, with eight living relatives with clear-cut bipolar disorder, type I, was a rich source of subjects. Were it absolutely clear that all eight had inherited this mood disorder from a common ancestor, this number might be sufficient roughly to localize the shared DNA segment or segments that carried the relevant alleles. But the chance of success would increase if more people who were also descendants of this ancestor could be studied. And even though the connection of any such people to Ana might go back too many generations to be remembered, there was reason to hope that by searching through church and civil records such distant relatives could be identified. Should they be found, they too might be members of families with a high concentration of manic-depressive illness.

THE LEAD TO very distant relatives of Ana's family who might share an affected ancestor came from the lithium clinic at Calderon Guardia Hospital. A review of records provided by Luis Meza and Alvaro Gallegos quickly yielded the identities of three additional families with a high concentration of mood disorders. Two of them, which I will call Families 2 and 3, would themselves have provided excellent material for this detective work; but both included relatives who were prominent public figures either in Costa Rica or in neighboring Panama or Nicaragua, and there was concern that their connection with the study might leak out. Despite assurances that details of their stories would be changed to conceal their identities (as has been done in the description of Ana's family), several members of Families 2 and 3 decided that the danger of disclosure was too great to justify their participation.

Fortunately, Luis and Alvaro knew of yet another group of affected relatives, Family 4, which has even more members with unmistakable bipolar disorder, type I, making it the most valuable of all; and members of this family were willing to participate. In their pedigree on the following page, only those with this diagnosis and a few with major depression are shown, whereas some other family members with evidence of a mood disorder were either not included in the diagram or are indicated as "unknown," to disguise the identity of this family further.

FAMILY 4

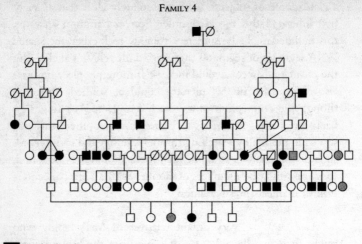

■ Bipolar disorder, type I
▨ Major depression
□ Unaffected or unknown

Then, to Eduardo Fournier's great delight, a search of their genealogical records turned up an eighteenth-century ancestor of Family 4 who was also an ancestor of Family 1. This finding raised the possibility that the same founder who may have introduced manic-depression into Family 4 might also be implicated in the mood disorders of Family 1—Ana's family.

Adding together Family 1 and Family 4 gave more than two dozen known relatives who shared a clear-cut diagnosis of bipolar disorder, type I. The next step would be to find out if they also shared one or more segments of DNA that were derived from a common ancestor.

8
HOT SPOTS
IN THE GENOME

Whenever you can, count.
   —Francis Galton

While the members of Ana's extended family were being evaluated clinically, great strides were being made in mapping human chromosomes. The starting material for this work was a set of standard DNA samples from several dozen families that had been collected in the 1980s at the Centre d'Etude du Polymorphisme Humain (CEPH) in Paris and distributed to geneticists around the world. Using the CEPH samples facilitated the relative positioning of newly discovered markers—mostly short tandem repeat polymorphisms (stirps)—on the emerging chromosome maps. These maps are continually updated through an international cooperative effort, as part of the Human Genome Project.

Much of the mapping is being done in France at a foundation called Généthon (the name—derived from "gene telethon"—reflects its support by the French equivalent of the Jerry Lewis telethons that helped fund the discovery of the gene responsible for muscular dystrophy). Another major mapper of stirps is a consortium of American scientists, many of whom are organized as the Cooperative Human Linkage Center. By mid-1994, when the genome screening of Ana's family was in full

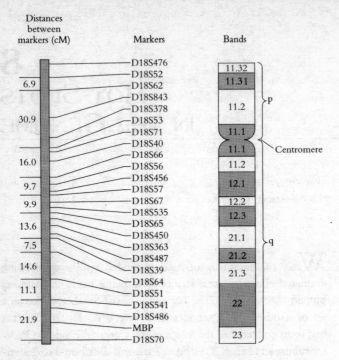

Distances between markers (cM)	Markers	Bands

swing, these French and American collaborators had localized about six thousand markers on the map, so the average distance between known markers was less than 1 centimorgan (roughly a million base pairs).

A typical chromosome map includes a drawing of its physical features, the locations of some named markers, and measurements of the distances between them. The figure above, for example, is a map of human chromosome 18 published in *Science* in 1994.

A critical physical feature of each chromosome is the location of a natural point of constriction, the centromere, which separates it into two segments called arms—a short one called p (for the French, *petit*) and a long one called q. Other features are brought out when the chromosome is stained with a dye that adheres to certain regions, giving rise to a distinctive pattern of bands (shown in alternating gray and white in the diagram). These bands have been assigned numbers relative to the centromere that serve as zip codes for markers and genes. Lined up next to the chromosome in the drawing are the positions of some stirp markers, which serve as more precise addresses. These markers may appear to be closely packed, but there are actually considerable intervals between them, indicated at the far left of the diagram. Though chromosome 18 is one of the smaller human chromosomes, it is still huge in molecular terms: it contains a double strand of DNA that is about 150,000,000 base pairs long and spans about 150 centimorgans, with about 25 genes in each centimorgan.

Similar maps are available for all the other chromosomes. There are also many additional markers that can be employed to explore a particular chromosome region in more detail. Both the basic maps and the continuous development of supplementary markers have proved to be indispensable tools in the hunt for mood genes in Ana's extended family.

THAT HUNT BEGAN toward the end of 1993 with a screen of all the chromosomes of each person with narrowly defined manic-depressive illness—that is, bipolar disorder, type I—in Costa Rican Families 1 and 4, using 307 markers based on stirps. What this meant in practice is that a group of researchers, including Alison McInnes, Michael Escamilla, Nelson Freimer, and their staff in San Francisco, and Pedro Léon and his staff in San José, began the tedious job of examining the DNA samples from more than two dozen manic-depressive people, as well as other relatives, with 307 pairs of custom-made bits of DNA— the primers—each designed to assess a particular polymorphism. The task comprised many thousands of laborious examinations, each carefully checked for accuracy.

Though measurements of stirps have now been automated, in 1993 they were done by hand—by going through a set of steps that began with the mixing of each DNA sample with each set of primers and ended with the careful measurement of the lengths of the stirps. And the sole immediate reward for all this work was the knowledge that every measurement was being done accurately. Only after the stirp lengths in each sample for every one of the 307 markers had been determined, transmitted to Lodewijk Sadkuijl in Holland, and analyzed by a complex computer program—an overall process that would take more than a year—would this large team of researchers learn if they had succeeded in finding any evidence of linkage. Unless they did, all their labors would have been for naught.

Fortunately, the initial screen identified promising regions on several chromosomes. These were then scrutinized further with an additional 166 markers (more thousands of elaborate measurements), each known to cover a region adjacent to one that showed promise in the initial screen. Were a promising marker truly close to a mood gene, nearby markers should also give evidence of linkage. If they didn't, that would argue against the possibility that this chromosome region contained a mood gene.

Of the 473 markers used in the two-stage procedure, 23 met the screening criteria for possible linkage to a mood gene in either Family 1 or Family 4, and 6 met the screening criteria for the combined data set from both families. They are indicated in the diagram on the following page by circles (Family 1), diamonds (Family 4), or stars (combined families)—with all the chromosomes depicted as the same size rather than drawn to scale.

Of the regions identified, the most likely site of a mood gene appears to be the cluster of three stars in the lower part of chromosome 18, the area with the broad zip code of 18q22-23; this cluster of identical marker alleles is shared by 23 of the 26 people with manic-depressive illness. What this clustering implies is that this chunk of chromosome 18—and the mood-gene allele it presumably contains—came from a common ancestor with manic-depression. And even though it was still possible that the apparent sharing of this chunk of chromosome 18 in relatives with the mood disorder was just a coincidence, statistical analyses of the results (using several methods) indicated that

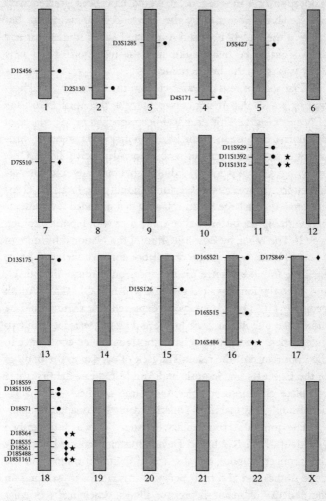

• Family 1　♦ Family 4　★ Combined families

this was unlikely. Five years of driving over the potholed roads of Costa Rica and of meticulous psychiatric and genetic testing had finally produced an encouraging result: 18q22-23 appears to contain a mood gene.

HAD THIS COSTA RICAN study been the first to report a likely location of a mood gene, it would have been greeted with considerable enthusiasm by the scientific community. But because of the widely heralded Amish and Israeli studies that had turned out to be disappointments, the reception of this new study was respectful but restrained.

That tone of restraint was set from the moment of publication by an accompanying commentary in the April 1996 issue of *Nature Genetics.* This essay had been solicited by the editors of the journal because the same issue also included two other chromosome screens of families with manic-depressive illness. One, by Douglas Blackwood of Edinburgh University and his collaborators, reported evidence for a mood-gene locus on chromosome 4p in a large Scottish family that included seven members with bipolar disorder, type I, and four with bipolar disorder, type II. The other, by Edward Ginns of the National Institute of Mental Health and a group of collaborators that included Janice Egeland, found evidence for hot spots on chromosomes 6, 13, and 15 in the long-awaited genome screen of Old Order Amish pedigree 110. Though everyone expected that there would be more than one mood-gene locus, finding the *same* locus in two independent studies would have greatly increased confidence in the importance of the result. The lack of overlap in the findings of the Costa Rican, Scottish, and Amish studies—the first three complete chromosome screens of large families with manic-depression—was therefore understandable but disappointing.

The journal's commentary, written by two distinguished geneticists, Neil Risch and David Botstein, reflected not only their concerns about the absence of common findings in the three studies, but also their personal frustrations in working on the genetics of manic-depressive illness. Risch had particular reason to be wary, because he had been a collaborator in Miron Baron's Israeli study, which presented strong initial evidence for a mood-gene locus on the X chromosome that was not confirmed by subsequent DNA testing (Nelson's crushing disappointment). Botstein, one of the originators of the idea of a human riflip map, was a collaborator in another study of families with manic-depressive illness that also suggested the existence of a mood gene on chromosome 18—but in a different region. Their commentary, titled "A Manic Depressive History," began:

> The mapping of inherited disease genes onto the human genome by linkage analysis has proceeded at an astonishing rate since the first such achievements in the early 1980s. Many hundreds of traits have been convincingly mapped, and positional cloning [that is, precise identification] of many scores of genes followed, often hard upon the heels of the linkage mapping. The same cannot be said, unfortunately, for traits that are inherited in a more complex fashion. Here the record is one of repeated claims for a variety of different loci followed by counterclaims and even retractions.
>
> In no field has the difficulty been more frustrating than in psychiatric genetics. Manic depression (bipolar illness) provides a typical case in point. Indeed, one might argue that the recent history of genetic linkage studies for this disease is rivaled only by the course of the illness itself. The euphoria of linkage findings being replaced by the dysphoria of nonreplication has become a regular pattern, creating a roller coaster-type existence for many psychiatric genetics practitioners as well as their interested observers.

Having made their point about the inherent difficulty of linkage studies of complex traits, Risch and Botstein went on to consider earlier attempts to locate mood genes, including two that had already implicated chromosome 18. The first was published in 1994 by a group led by Wade Berrettini of Jefferson Medical College and Elliot Gershon of the National Institute of Mental Health. Their patients came from twenty-four different families and included people with broadly defined manic-depressive illness (bipolar disorder, type I *and* type II). Having begun with the intention of screening all chromosomes, they

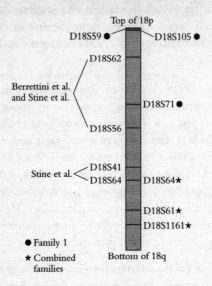

Top of 18p

D18S59 ● D18S105 ●

D18S62

Berrettini et al.
and Stine et al.

D18S71 ●

D18S56

Stine et al. < D18S41
D18S64

D18S64 ★

D18S61 ★

D18S1161 ★

● Family 1
★ Combined
families

Bottom of 18q

soon focused on eleven markers that covered much of 18p and
part of 18q, and found some evidence of linkage with five of
them. Stimulated by this work, a second group, including Colin
Stine and Raymond DePaulo of Johns Hopkins University
as well as Botstein, also examined chromosome 18. Based on
studies of patients from twenty-eight families with broadly defined manic-depression, their 1995 publication provided some
support for Berrettini and Gershon's findings. It also presented
evidence for a locus in the lower part of chromosome 18q, as
shown in the diagram above.

Also in the diagram are the Costa Rican results for the combined families (stars) and for Family 1 (circles). Instead of providing confirmation of the earlier reports, the Costa Rican
studies extended the possible locations of mood genes to virtually the entire length of chromosome 18. The extension to the
lower portion of chromosome 18 was supported by two other
studies, also published in 1996, one by a Belgian group led by

Christine Van Broeckhoven and the other by a Utah group led by William Byerly.

What is the upshot of these five studies of families with manic-depressive illness that all point to chromosome 18? Does chromosome 18 really contain a mood-gene locus? More than one? And if it does, where exactly are they?

TO ANSWER THESE questions with respect to the Costa Rican families, several approaches are being taken. The first is to try to increase the size of the patient sample by persuading additional relatives of members of Families 1 and 4 to participate in the study. The second is to examine DNA samples from the relatives with other mood disorders (bipolar disorder, type II, and recurrent major depression) who were excluded from the initial study, to see whether or not they too share these hot spots. In addition, Mitzi Spesny and others remain in close contact with the core members of these families, looking for new cases of manic-depression among the large number of adolescents who are reaching the age of maximal risk. As data accumulate on additional members of these families, linkage to the hot spots already identified will be either confirmed or refuted. And as more is learned, through the Human Genome Project, about the genes that are located in 18q22-23, it will become increasingly easy to determine if there is a consistent structural variation in one of these genes in people with manic-depression. Such a finding would help to implicate that gene as a mood gene.

At the same time a very different approach is also being taken to hunt for mood-gene loci in the Costa Rican population. It is based on the assumption that in a population descended from a small founding group, with little interbreeding with outsiders, the allele (or alleles) responsible for a particular phenotype (such as a disease) can be traced to just a few ancestors. If the disease is very unusual (such as Huntington's disease) and the founding population is sufficiently small, it might even be traceable to a single founder. With more common diseases (such as manic-depressive illness) it would be more likely that there would be several founders—but each lineage would still be traceable by a combination of genealogical records and DNA testing.

Because Costa Rica's Central Valley was settled by only eighty-six Spanish families—a few hundred people—and because manic-depressive illness (that is, bipolar disorder, type I) affects only about one person in a hundred, it can be argued that just a few founders were the source of much or all the manic-depression in the settlers' descendants. To simplify the argument, let us start with the assumption that all descendants of the Central Valley founders who develop manic-depressive illness inherited a critical mood-gene allele from just a single founder. Were this assumption true, the founder would transmit not only this mood-gene allele to all descendants with manic-depression, but also the whole set of surrounding marker alleles that happened to be on the chromosome where that mood-gene allele resided. Being such close neighbors, these marker alleles would not (on the average) have separated from the mood-gene locus by crossing over through more than a dozen generations. If this logical chain is based on a true assumption, it should be possible to find the mood-gene locus by looking for a shared bit of a chromosome with an identical cluster of marker alleles in all the people with manic-depressive illness—even if these people have no knowledge that they are actually distant relatives. *No need to look for families:* anyone with manic-depression who is descended from the original founding population *would have to be a distant relative,* since there would have been only one source of the bit of DNA that brought the disease to the Central Valley.

Parallel reasoning enabled Pedro Léon and his collaborators to identify first the locus and then the gene for a hereditary form of deafness, which he eventually traced to Felix Monge and his Spanish ancestors. But unlike the hereditary deafness Pedro was studying, which is rare and caused by a single gene, manic-depression is neither rare nor monogenic. This makes it hard to believe that just one mood-gene allele derived from a single founder could be the main source of all (or even a large part of) the manic-depression in the descendants of the Spanish founders who settled in the Central Valley. There is also the matter of the twelve hundred Amerindian ancestors of the Central Valley population who, about 1700, began intermarrying with the somewhat larger number of people presumed to be of pure Spanish origin. Since the Amerindians would be expected to have had some manic-depressive illness, they too should be a source of mood-gene alleles that increase susceptibility to this disorder. Taking these problems into consideration, I for one was skeptical that the approach Pedro used would work for manic-depression.

But Michael Escamilla decided to try it. To increase his chances of success he has confined his attention to people with the most severe forms of manic-depressive illness, requiring for inclusion in the study a minimum of two hospitalizations for florid symptoms. And to identify descendants of the original Central Valley population he has set a criterion that most of the great-grandparents of those studied (preferably all eight) were themselves born in the Central Valley, which remained quite isolated until the beginning of the twentieth century.

So far Michael Escamilla has identified more than a hundred people with florid manic-depressive illness who can be traced back to Central Valley settlers, and he and Alison Mc-Innes have studied in great detail the DNA of forty-eight of them, concentrating on chromosome 18. The main finding of this experiment of nature is that a substantial fraction of the patients with this very severe form of manic-depressive illness share several marker alleles that are closely packed in the region of 18p11.32 (the region that also showed maximal evidence of linkage in Family 1, Ana's family)—suggesting that these marker alleles were indeed derived from a single founder. This conclusion is also supported by Eduardo Fournier's ongoing genealogical work, in which he is searching through civil and church records for evidence of shared ancestry among the Central Valley patients with severe manic-depression. The

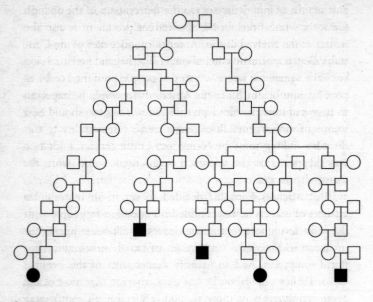

●, ■ Bipolar disorder, type I

diagram above summarizes evidence Eduardo found that five of these patients—who did not know they were related—have common ancestry.

These five distant relatives with manic-depression also share a cluster of marker alleles of 18p11.32, which includes one allele found in Ana's family. Furthermore, the region of chromosome 18 shared by these five affected people (and many others) is quite small, just as would be expected if it had been passed intact through all the generations (and crossings over) that separate the affected individuals from the presumed founder. It is, in fact, small enough that genes known to be in that region are already being examined for mutations that could lead to their identification as mood genes.

ENCOURAGING THOUGH THESE findings are, much more work is needed to establish the precise genetic basis of manic-depressive illness. Even if the Costa Rican project pays off soon, the pattern of inheritance of manic-depression and the multiple leads that have been found by different groups of researchers indicate that there will still be other mood genes to find, and complicated interactions with environmental and genetic backgrounds to grapple with. We can get some idea of the complexity of the problems that lie ahead, and the promise of their solution, by considering the much more advanced genetic studies of another serious mental illness—Alzheimer's disease.

One indication of the complexity of the genetics of Alzheimer's disease is that most cases, though known to be influenced by genes, are not concentrated in particular families. Many of those which are obviously familial have been traced to dominant alleles of three causative genes. In addition, a susceptibility gene has been found that increases the risk of developing the disease at a somewhat younger age, and in particular circumstances.

Though so readily summarized, these remarkable findings did not come easily. Instead, as with manic-depressive illness, there were years of controversy and of seemingly contradictory findings, which can now be explained. Some of the confusion was caused by lumping together cases of early-onset Alzheimer's disease (symptoms appearing before the age of sixty, often before fifty) and cases with late-onset disease (symptoms appearing after the age of sixty, often in the seventies, eighties, or later). Only when they were considered separately did it become clear that early-onset and late-onset cases of Alzheimer's disease were influenced by different genes. Further confusion was caused by lumping together early-onset cases from different families that did not all have the same genetic basis (another example of the problem posed by genetic heterogeneity, which we first encountered in considering the different origins of wrinkled peas). What helped clear up the confusion was the decision to do some separate linkage studies on large extended families derived from distinct populations, as is being done in the current work on manic-depressive illness in Costa Rica.

Of these distinct populations, a notable one is the "Volga Germans," so called because they are derived from about thirty thousand people who emigrated from Germany to the Volga valley in the eighteenth century. They settled there at the invitation of Catherine the Great, herself of German origin, and did not mix with the native Russians. Among their descendants are a number of families with an inherited form of Alzheimer's disease that tends to appear before age fifty, all likely to be traceable to a single founder. As long as the Volga German families were lumped together with other families with different backgrounds, it was not possible to identify a shared Alzheimer's disease locus in the combined group that resulted—because there wasn't one! Only when the Volga German group was analyzed separately did it become apparent that the locus responsible for their disease resided on chromosome 1. This finding led to the identification of the causative gene named *PS2*, which encodes a protein that has been named presenilin-2.

We also now know that there are other sets of families with early-onset Alzheimer's disease attributable to different causative genes whose identities have been clearly established. One of these, *PS1*, on chromosome 14, encodes a protein called presenilin-1. The other, *APP*, is on chromosome 21 and encodes a protein called the amyloid precursor protein. Unlike the presenilins, whose functions have not yet been determined, we now know that a fragment of the amyloid precursor protein is a critical component of the abnormal protein deposits called amyloid bodies that Alois Alzheimer originally observed in the seminal autopsy of the shrunken brain of a woman with progressive dementia that he performed about ninety years ago.

STUNNING AS THESE discoveries are, they explain only the small fraction of cases of Alzheimer's disease with early onset and a clear-cut pattern of inheritance. In the vast majority of cases, the symptoms tend to begin after the age of sixty, and there is usually no obvious pattern of inheritance. Nevertheless, we already know a great deal about one of the susceptibility genes for the late-onset form of Alzheimer's disease that has a substantial effect even in many people who don't have any evidence of this disease in their families.

Identification of this gene has had an enormous impact because the late-onset form of Alzheimer's disease directly concerns us all. Most experts believe that—if we live long enough—every one of us will inevitably develop the brain degeneration and dementia that are the hallmarks of this disease. But some will show signs of dementia at the relatively early age of sixty, while others may be spared for decades and die from other causes with their minds intact. The marked variation in the age when dementia begins is controlled, in part, by alleles of susceptibility genes.

The first hint of a gene that influences susceptibility to Alzheimer's disease came from a linkage study of eighty-seven affected people from thirty-two families by Allen Roses, Margaret Pericak-Vance, and other scientists at Duke University. Using only thirty-six markers that covered parts of only eleven chromosomes, they were lucky enough to detect a hot spot on chromosome 19. Recognizing that the result was far from conclusive, Roses and Pericak-Vance presented it in 1991 as a preliminary finding; others viewed it with even greater skepticism.

This tentative finding became a major breakthrough in understanding the genetics of Alzheimer's disease in a manner reminiscent of Pauling's work with sickle cell anemia. Just as Pauling's great advance was based on substantial knowledge about hemoglobin, the Alzheimer's breakthrough was based on substantial knowledge about a protein called apolipoprotein E (APOE) that influences the amount of cholesterol in the blood, thereby affecting the risk of heart attacks. Researchers had already shown that the protein exists in three forms, APOE2, APOE3, and APOE4, each encoded by a different allele of the *APOE* gene, and that APOE4 is associated with high levels of cholesterol whereas APOE2 is associated with low levels of cholesterol. To learn more about the alternative forms of this protein, they went on to identify the *APOE* gene; and, in the course of its identification the location of the *APOE* gene was mapped to the same region of chromosome 19 to which Roses and Pericak-Vance would localize a hot spot for Alzheimer's disease.

But finding some evidence of linkage of Alzheimer's disease to a region of chromosome 19 that happened to contain the *APOE* gene—among hundreds of others—was hardly sufficient to make the connection with that particular gene. In advance there was no reason to suspect that APOE, a protein that influences blood cholesterol, would have anything to do with the brain cell degeneration found in Alzheimer's disease.

The connection was made from a study of amyloid, the substance that Alzheimer had seen in his famous autopsy. Because amyloid seemed to play a critical role in Alzheimer's disease, researchers became interested in identifying proteins that could bind to amyloid, since they too might be involved in the disease process. Of the many proteins in the brain that bind amyloid, one turned out to be APOE. But no one paid any attention to this observation until it was noticed that the Alzheimer's disease hot spot on chromosome 19 was in the same region as the *APOE* gene. Then bells started ringing. Was it possible that the linkage study that had pointed to chromosome 19 was actually pointing to the *APOE* gene? Might the three known alleles of *APOE*—*APOE2, APOE3,* and *APOE4*—have different effects on susceptibility to Alzheimer's disease? Could *APOE,* which was already known to be a heart attack susceptibility gene, also be an Alzheimer's disease susceptibility gene?

To answer these questions, the frequencies of the three known alleles of *APOE* in people with Alzheimer's disease and in the general population were compared. Having one *APOE4* allele, it turned out, increased the risk of developing Alzheimer's disease; having two *APOE4* alleles (genotype *APOE4,4,* found in about 2 percent of the population the Duke group initially studied) increased the risk even more. To indicate the relative risk for people with various *APOE* genotypes at different ages, in 1996 Roses published the diagram on the facing page. Because Roses is among those who believe that all of us will develop Alzheimer's disease if we live long enough (in his view, by the age of 140), he drew the graph in a form that shows the diminishing percentage of people with each genotype who would be expected *still to be free of this disease* at increasing ages.

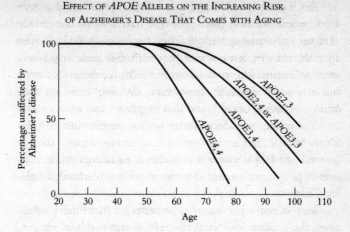

EFFECT OF *APOE* ALLELES ON THE INCREASING RISK
OF ALZHEIMER'S DISEASE THAT COMES WITH AGING

Included in this diagram is the startling result that having the *APOE4,4* genotype implies that you have more than a fifty–fifty chance of showing signs of brain degeneration by age eighty (or less than a fifty–fifty chance of being unaffected).

Should this diagram impel you to try to find out about your personal vulnerability to Alzheimer's disease by having your *APOE* genotype determined, you may be surprised to learn that the DNA test is unavailable to you. Unless you have sufficient symptoms of dementia for your doctor to suspect that you may already have Alzheimer's disease, the company that markets this test, Athena Neurosciences, won't examine your DNA. Their decision to offer the test only as an aid in making the diagnosis of Alzheimer's disease in people with symptoms of dementia reflects the consensus of a large group of experts that *APOE* genetic testing should not presently be used for predictive testing of asymptomatic people because "it does not foretell disease."

The twenty researchers and scholars who reached this consensus, which was published in the *Journal of the American Medical Association* in March 1997, are experts in genetics, genetic counseling, ethics, and public policy; Roses is among them. Though they agree that having an *APOE4* allele increases the risk of developing Alzheimer's disease, and that having the

APOE4, 4 genotype increases it greatly, they point out that even those people with the latter genotype have a significant chance of never becoming demented. They fear that "individuals who discover that they have at least [one *APOE4*] allele may falsely attribute normal forgetfulness to the onset of Alzheimer's disease and may make significant life-altering decisions based on such misinterpretation of risk"; and that employers and health insurance companies will discriminate against people with *APOE4* alleles. Similar fears are sure to arise when mood genes are discovered and DNA testing is considered for asymptomatic relatives of people with manic-depression, such as Michael's daughter, Charlotte.

Were effective preventive treatments for Alzheimer's developed, the experts who want the *APOE* test withheld are prepared to change their minds. Though they are not yet willing to make any exceptions to their ban on predictive testing, they already acknowledge that knowing your *APOE* genotype may have practical value in certain circumstances. For example, a recent study suggests that having an *APOE4* allele increases vulnerability to long-term brain damage from sports such as boxing that are likely to expose you to multiple blows to the head. Should this be substantiated, it would be a dramatic example of the development of a disorder through the interaction of a specific allele of a susceptibility gene with a clearly defined environmental factor. It might also justify *APOE* testing for people considering activities that expose them to the risk of repeated head trauma.

THE GREAT RECENT progress on Alzheimer's disease, a disorder that once seemed to defy genetic analysis, serves as a model for research on manic-depression. Just as some rare forms of Alzheimer's disease can now be attributed to single causative genes, so too may there be rare forms of manic-depression in which alleles of a single gene play a central role. And just as most cases of Alzheimer's disease are attributable to interactions of alleles of susceptibility genes with a person's genetic and environmental background, so too are most cases of the mood disorder likely to reflect interactions between such background factors and certain alleles of mood genes.

Which raises a very puzzling question: why might the human genetic repertoire contain fairly common mood-gene alleles that predispose so many people to manic-depression?

9

BOTH GOOD SEED
AND BAD

*. . . to our chemical individualities are due our chemical merits as well as our
chemical shortcomings. . .*

—Archibald Garrod (1931)

*I have often asked myself whether, given the choice, I would choose to have
manic-depressive illness. If lithium were not available to me, the answer would
be a simple no—and it would be an answer laced with terror. But lithium
does work for me, and therefore I suppose I can afford to pose the question.
Strangely enough I think I would choose it. . . . Because I honestly believe
that as a result of it I have felt things, more deeply; had more experiences, more
intensely. . . . And I think much of this is related to my illness—the intensity it gives to things and the perspective it forces on me . . .*

—Kay Redfield Jamison (1995)

The widely held view of genetic diseases is that they are caused
by "bad seeds"—alleles that lead to disability and suffering and
even death. But this simple view does not hold for all alleles that
contribute to or cause genetic diseases. In many instances such
alleles may also confer substantial benefits in particular circumstances. Should they have a positive net effect on the reproductive success of those who carry them, more and more children
would be born with these alleles. They might, in fact, become
quite common—and some of them do.

As studies of families with manic-depressive illness continue, the realization is growing that alleles which increase vulnerability to this disorder may actually have favorable effects on aspects of temperament—such as charm, enthusiasm, and sexuality—that may compensate for the irrationality and even suicide of the extreme form of this disease. Positive features may even be the only forms of expression of these mood-gene alleles in some members of these families, enhancing reproductive success and accounting for their considerable prevalence in the population. Such favorable features may in fact have led to the evolution of many different kinds of mood-gene alleles, making it difficult for geneticists to pick out any single one of them.

The idea that there may be favorable effects of the very same mood-gene alleles that increase vulnerability to manic-depression is mere speculation for now. But it is an idea that is supported by many beautifully documented examples of the evolution of genetic variations that influence other traits. Consider, for example, the allele that causes sickle cell anemia.

SICKLE CELL ANEMIA, the object of Pauling's famous hunch, is transmitted in the same Mendelian recessive pattern that I described for AKU. People with one sickle cell allele (s) and one normal allele (S) have no obvious deleterious effects, while those with two sickle cell alleles (genotype ss) develop symptoms. What makes this such a compelling example is that *about three out of ten* adults in certain parts of Africa carry this potentially fatal allele. Why has the sickle cell allele become so prevalent?

The unavoidable conclusion is that the sickle cell allele must—at least at some point in history—have conferred an advantage so great that it outweighed the lethal threat that it also conveys. But what could be the advantage of such a dangerous allele?

The answer was provided in 1954 by Anthony Allison, a British physician who had been raised in Africa and who had been infected with malaria as a child. While studying the distribution of blood types among Africans, Allison noticed that sickling of red blood cells was much more common among tribes that lived in wetlands where mosquitos bred and transmitted malaria. From this and other evidence he concluded not only that sickle cell anemia is somehow related to malaria, but also that a single sickle cell allele may, in some way, protect those people who inherit it against this parasitic disease. But how does a sickle cell allele confer such protection?

We now know that the sickle cell allele does this by encoding an alternative form of beta-globin, thereby changing certain properties of infected red blood cells so that both the cells and the parasites within them are destroyed. Strong evidence for the effectiveness of this antimalarial effect comes from studies in Nigeria, where a particularly virulent form of malaria is prevalent. In this population 24 percent of newborns have a sickle cell allele, but in five-year-olds the proportion has risen to 29 percent—because of the death from malaria of more than 5 percent of the youngsters who lack the protection that this allele provides.

A major factor that has favored the selection of the sickle cell allele is that people with only one copy of *s* (and one copy of the normal allele, *S*, that is, with genotype *Ss*) *pay no price* for the benefit of its antimalarial effect. Except, of course, if they marry others who also are *Ss*. In that event, one in four of their children are expected to be *ss*, and afflicted with a serious disease. Their consolation for this tragedy is that half of their children would be expected to be *Ss* and therefore reap the antimalarial benefit of a single sickle cell allele, with no signs of illness; only one in four of their children would be *SS* and thus unprotected.

So the sickle cell allele is both good seed and bad, a mixed blessing that brings advantages and disadvantages to individuals, families, and the population as a whole. In Nigeria an evolutionary balance has been reached between the curse of sickle cell anemia and the blessing of protection against fatal malaria: each generation is born with about 15 percent sickle cell alleles and 85 percent normal alleles. With these allele frequencies and random mating, more than a quarter of the people are born protected (that is, *Ss*) at a cost of about 2 percent who are born *ss* and have sickle cell anemia. Were the frequency of the sickle cell allele to increase much above these levels, the probability of sickness due to *ss* would rise steeply, whereas a lower frequency would apparently give too little protection to the population.

The sickle cell allele is not the only human genetic counter-attack against the parasite that causes malaria. Among the others is a group of hundreds of different mutations that each reduce the levels of one or more of the protein components—globin chains—of hemoglobin, and that together are the cause of the commonest human genetic disease, the form of anemia called thalassemia (which Gerhard and Housman were studying when they decided to work on manic-depression in the Amish). As with sickle cell anemia, inheritance of a single copy of one of these abnormal alleles helps fend off malaria—by making the internal environment of red blood cells inhospitable to the parasite without any obvious detriment to the person—whereas inheritance of two abnormal alleles in various combinations gives rise to anemia and other symptoms.

This is not the end of the list of genetic defenses against malaria that are also mixed blessings. Yet another set of some-what harmful mutations has become prevalent because they too have antimalarial effects. This set of mutations—more than four hundred different ones are known—all diminish the activity of an enzyme called glucose-6-phosphate dehydrogenase (G6PD)—the very same enzyme that was used as a marker in the Israeli study of manic-depression (because it is so commonly mutated in Israelis whose ancestors lived in malaria-infested areas). These G6PD alleles also serve to illustrate another critical point: the importance of specific environmental factors in the development of symptoms of a hereditary disease.

The mutations in the gene that encodes G6PD exert their antimalarial effect by interfering with the manufacture of glutathione, an antioxidant that protects the internal environment of red blood cells. As a result, oxidants rise to toxic levels in red blood cells that harbor malaria parasites, leading to their destruction. But the downside of the lower antioxidant level is that it increases vulnerability to a particular environmental factor—oxidants in the diet.

The result is a disease called favism, which attacks vulnerable people after they have eaten fava beans, a Mediterranean delicacy rich in oxidants. These beans, which cause no harm to people with normal levels of G6PD, can cause massive destruction of red blood cells in people with the enzyme deficiency. Because about 25 percent of Mediterranean peoples have a G6PD deficiency (reflecting the fact that the region was once riddled with malaria), it is easy to understand why the Greek mathematician Pythagoras warned his disciples against eating fava beans. And the observation that only some Romans were vulnerable to favism, while others could enjoy fava beans with impunity, may account for the famous comment by the philosopher Lucretius (in his poem *On the Nature of Things*) that "what is food for one man may be fierce poison to others."

THE VARIOUS GENETIC reactions to malaria show that many different mutations can become prevalent in response to a dangerous environmental condition. But what exactly are the conditions that led to the proliferation of variations in mood genes? Not knowing the identity of mood genes, we can only speculate about what these conditions might be. Fortunately, such speculations can draw upon theoretical studies of the evolution of alternative behavioral strategies such as aggressiveness or passivity, which are correlated, respectively with mania or depression. In these theoretical studies the provocative condition is not an invasive parasite such as the one that causes malaria but the behavior of another member of one's own species—in fact, of one's own social group.

One example, drawn from game theory, has been popularized by the evolutionary biologist John Maynard Smith and is called the hawk–dove game. In this scenario, which is concerned with the evolution of fighting behavior, some individuals are genetically programmed to be aggressive (hawks) and others to be passive (doves), and the game provides some advantages and disadvantages to each strategy. Thus a hawk does well against a dove, who surrenders resources and retreats before being injured. But the hawk pays a price for its aggressiveness: should it encounter another hawk, the pair will engage in an escalating fight, each risking serious injury. These injuries, in turn, open the way for doves, who can share the resources no longer controlled by the disabled hawks. The net result is a mixed population with some hawks and some doves in a ratio that is determined by the balance of gains and losses from each strategy—a situation reminiscent of the ratio of phenotypes that reflect the balance of consequences of S and s alleles in malaria-infested Africa.

The behavioral strategies of Maynard Smith's hawks and doves are theoretical, and not based on real genes and alleles like those that we hope to identify in the study of mood disorders. But the gap between armchair theories about the evolution of behavior and direct examination of DNA variations is beginning to be bridged. A notable example is the recent study by Barry Sinervo and C. M. Lively, biologists from Indiana University, of the behavior of a small iguanid lizard (*Uta stansburiana*) that lives in the Coast Range of California. This study of male territorial and sexual behavior provides a real-life illustration of game theory and also raises the possibility of actually finding the relevant genes.

In this species of lizard, males can be divided into three categories: the very aggressive, who defend large territories and mate with all females in their extensive domains; the moderately aggressive, who defend much smaller territories and mate with and closely guard the sole female who lives there; and the "sneakers," who mate with females that are left unprotected but who do not defend any territories. What makes the behavior of these lizards relatively easy to study is that experimenters can distinguish the three types on the basis of the color of their throats: orange-throated males control the large territories, blue-throated males the smaller ones, and the non-territorial sneakers have yellow stripes on their throats (just like many of the females). Furthermore, both the throat color and the territoriality are inherited traits, and a new generation matures every year.

Why are there three types of males? Wouldn't the orange males simply take over? After all, they control lots of females, so they should become more and more abundant. Yet the persistence of the blues and the yellows indicates that they too must each have some selective advantage. In fact, what Sinervo and Lively found by observing these lizards over a six-year period in their natural habitat, was a marked fluctuation in the relative number of males of each type, which never reached a constant ratio. Furthermore, the fluctuation followed a clear pattern: a preponderance of blue (in 1991), then orange (in 1992), then yellow (in 1993 and 1994), then back to blue (in 1995).

From this, Sinervo and Lively concluded that as soon as one form began to predominate it opened the way to the particular alternative form that could defeat it. For example, as the orange males became more common and spent their energies fending each other off, sneakers could invade and mate with the females. Populations of the yellow sneakers could, in turn, be invaded by the blue-throats, who could successfully defend a single female against them. As the blue throats became more common, the stage was set for the reemergence of the orange-throated males, who were more than a match for them.

Following prior theoretical work by John Maynard Smith and others, Sinervo and Lively have likened this situation to the children's game called rock–paper–scissors. In their words:

> As in the game where paper beats rock, scissors beat paper, and rock beats scissors, the wide ranging "ultradominant" strategy of orange males is defeated by the "sneaker" strategy of yellow males, which is in turn defeated by the mate-guarding strategy of the blue males; the orange strategy defeats the blue strategy to complete the dynamic cycle.

A particularly interesting feature of this battle of the lizards is that even though the genetic basis for their personality differences has not been worked out, their ability to switch from a preponderance of one form to a preponderance of another form in just a single generation suggests to Sinervo that such complicated behavioral strategy may be controlled by alleles of one gene (or a very small number of genes). Among the candidates for this role are alleles of a gene that determines the animal's level of testosterone, the male sex hormone. Since each allele confers a reproductive advantage in the proper social circumstances, this assures their persistence in the genetic repertoire of these lizards.

In many other situations the balance between behavioral strategies doesn't oscillate from year to year. Instead the result is what Maynard Smith calls an "evolutionary stable strategy." A constant fraction of individuals adopts one strategy (for example, hawk) and another fraction adopts an alternative strategy (dove).

Considering the enormous influence of behavioral strategies on reproductive fitness, it is likely that, in the human world, alleles of many different genes that can affect them have become fairly prevalent because they have favorable effects in specific situations. And this may account for the proliferation of many different alleles of mood genes that can give rise to adaptive behavioral features in some combinations, but to manic-depressive illness in others.

IT IS, IN fact, easy to make the case that the milder form of mania, called hypomania, has many adaptive aspects. With it comes optimism, enthusiasm, charisma, confidence, boldness, decisiveness, risk-taking, and the uninhibited thinking that sometimes leads to creative ideas. These are attributes that are not only useful to the individual but also attractive to others, ensuring social position and reproductive success. Tempering mild mania with mild depression may also be useful, because the pain of depression may abort excessively exuberant conduct that has begun to lead to trouble; and the pain of depression may also serve as a signal to others, calling forth compassion and help. So even though some people with florid manic-depressive illness might be at a reproductive disadvantage, the benefits that certain of their relatives enjoy from the same mood-gene alleles that are less fully expressed may compensate sufficiently to maintain a high frequency of these alleles in the population—just as the benefits of one copy of the sickle cell allele balance out the serious illness of those with two copies.

And even people with full-blown episodes of mania and depression may also gain some benefits from their illness, as was already apparent in the case of Michael's mother, Flora. A leading advocate of this view, Kay Redfield Jamison, herself afflicted with manic-depressive illness that is now controlled with lithium, has written a book—*Touched with Fire: Manic-Depressive Illness and the Artistic Temperament*—in which she describes the enhancement of imaginative thinking during mild manic states. For example, the coming together of disparate ideas and images along with the sense of freedom from conventional restraint— very common experiences during the manic states of great artists who have manic-depression—can be translated into the vivid metaphors of the poet (Byron), the dazzling images of the painter (van Gogh), and the passionate melodies of the composer (Robert Schumann). Although artists with mood disorders often complain bitterly about their episodes of depression, they tend to take for granted, as their welcome birthright, the mild mania that so often comes with it.

Mood disorders are also commonly observed among creative people in other fields, such as science. A notable example is the most highly regarded of all scientists, Isaac Newton, whose well-documented mood swings are widely believed to be indicative of manic-depression. Among Newton's most severe episodes was a prolonged depression with thoughts of suicide in 1662, at age twenty, and another in 1693 that included paranoid delusions about his famous philosopher friend John Locke ("you endeavoured to embroil me with women"). Newton also had periods of exuberance and lavish spending, and a highly provocative irritability, which are all very common among people with manic-depression. The surprising lapses of rational thought of this exceptionally rigorous scientist were recorded by another famous friend, the essayist Samuel Pepys:

> I had lately [in 1693] received a letter from him so surprising to me from the inconsistency of every part of it, as to be put in great disorder by it, from the concernment I have for him, lest it should arise from that which of all mankind I should least dread from him and most lament for, I mean a discomposure in head, or mind, or both.

Lapses of rational thought and other signs of manic-depression are also well known among captains of industry. Consider, for example, these excerpts from a 1996 article in *The Wall Street Journal* about mood disorders among corporate leaders:

> It may not take a fruitcake to attempt a business empire, but many who do are fruitcakes. Robert Campeau, despite a history of mental instability and periods of babbling incoherence, was able to borrow enough money to buy and wreck most of the U.S. department store industry. Another Canadian, Pierre Peladeau of giant Quebecor, has acknowledged episodes of erratic judgment until he got his moods under control with lithium. His revelations were featured in *Canadian Business* magazine, which argued that manic-depressive disorder is so widespread in corporate life that it could be called "CEO's disease." . . .
>
> Now the most interesting question: Many corporate maniacs later attribute their brilliant careers to their mental illness, but are they right? John Mulheren, before he was arrested driving toward Ivan Boesky's house with a small arsenal in his trunk, built up his arbitrage firm, Jamie Securities, with frenetic energy. Later he was quoted saying his manic-depression was "the only reason I've been successful."
>
> And even those who put themselves under a doctor's care often choose to go off their medication when facing an important deal or deadline, believing their illness gives them

a creative boost. It's noteworthy that Ted Turner had been on lithium for more than a decade, ever since almost spending his empire into the ground. Now he claims he was misdiagnosed from the start, but it's also the case that he stopped taking the drug while the epochal merger of Turner Broadcasting and Time Warner was being hammered out. . . .

RECOGNIZING THAT PEOPLE with manic-depressive illness can be enormously creative and productive doesn't relieve us, however, of the responsibility of trying to find the genetic underpinnings of their disorder. People with this illness usually have their lives disrupted by it. Many wind up in jail or with ruined careers because of embezzlement or fraud when episodes of mania make them feel that they are above the law. All consider suicide when deeply depressed, many are driven to attempt it, and almost one out of five people with full-blown manic-depression succeed in killing themselves. Although drugs and psychological treatments are available to prevent attacks and to temper symptoms, these are often ineffective. And even when a drug like lithium seems to be working, many stop taking it because they miss the euphoria and creativity that comes with mild mania, only to find themselves once again out of control. Should you ask people with manic-depression or their relatives what they think about this illness, few would argue that its benefits outweigh its disadvantages. In fact, as we shall now see, many are so distressed by its ruinous effects that they are extremely worried about the danger of passing it on to the next generation.

10
GRAPPLING WITH FATE

. . . the stated intention of much genetic research is eugenicist by implication. Implicit in every research grant written for the study of a genetic disorder is the suggestion that the disorder may be corrected or that identification of a "causative" gene or genes will help in population screening or fetal diagnosis.

—Editorial, *Nature Genetics* (1997)

The biggest practical consequence of the widely publicized hunt for mood genes is a heightened awareness that there is an inherited vulnerability to manic-depressive illness. As a result, many people like Michael or Ana who have relatives with manic-depression are concerned about what this means to themselves and their families. Are they, or their children, also fated to develop a mood disorder?

Were manic-depressive illness a simple Mendelian disorder with a known genetic basis—like Huntington's disease—this question could be answered very precisely on a case-by-case basis. All that would be needed is a sample of a person's DNA, which could be directly examined to determine whether or not the abnormal allele had been inherited. If it had been, the disease could be confidently expected, provided the person lived long enough for it to make its appearance. If it had not been, there would be no risk of developing the disease.

But manic-depressive illness is not a simple Mendelian disorder. Instead, like many other prevalent diseases such as diabetes, it is a complex disorder that appears only through the combined effects of alleles of susceptibility genes and environmental factors. So even when mood genes are identified, genetic testing will not have the same predictive value that it does for Huntington's disease. We already know that if two people have *exactly the same* combinations of mood-gene alleles—as is the case with identical twins—this doesn't guarantee that if one develops manic-depression the other will as well.

Despite the fact that manic-depression is not a simple inherited disorder, people who have relatives with this disease want some idea of the chances that they or their children will also be stricken. The estimate they can be offered, called an empirical risk estimate, is based on the systematic studies of large numbers of affected families we looked at in Chapter 2, as well as other summaries of family studies. Because the studies included in these summaries did not all use identical diagnostic criteria, and must surely include people with mood disorders of different origins, the following risk estimates should be viewed as no more than approximations:

1. If you have a first-degree relative (a parent or a sibling) with narrowly defined manic-depressive illness (that is, bipolar disorder, type I) you have about an 8 percent lifetime risk of developing bipolar disorder—8 chances out of 100 of developing this disorder at some time in your life—which is roughly ten times the risk to the general population. But you have 92 chances out of 100 of never coming down with it. You also have about a 10 percent lifetime risk of having an episode of major depression, about twice the risk to the general population.

2. In families with an unusually large percentage of affected relatives (as in Costa Rican Families 1 and 4) and unusually severe cases of bipolar disorder, the lifetime risk of bipolar disorder to first-degree relatives is somewhat greater than 8 percent.

3. If you have a first-degree relative with major depression, you have about a 10 percent lifetime risk of having an episode of major depression (twice the risk to the general population). If a first-degree relative became depressed before the age of twenty, your lifetime risk of having an episode of major depression goes up substantially, to about 30 percent. But this also means there are almost 3 chances in 4 that you will be spared.

4. If you have one second-degree relative, such as an uncle, aunt, or grandparent, with bipolar disorder, and no affected first-degree relative, your lifetime risk of developing a major mood disorder is not much greater than the risk to the general population (about 1 percent for bipolar disorder and 5 percent for major depression).

Some people in affected families breathe a sigh of relief when they learn about these estimated risks, others find them alarming. In either case, statistics like these should be regarded with caution in making personal decisions like whom to marry or whether to have children. Discussion with a psychiatrist interested in genetics or a genetics counselor experienced in working with families with mental illness can be very helpful for a number of reasons.

For one thing, it's important to know if the relatives in question really have manic-depressive illness, and trained personnel can help to assemble the facts about a particular case to answer that question. Often the diagnosis turns out to have been incorrect, to have been improperly communicated, or to have been misunderstood.

Discussion with experts in the field is also helpful in assessing individual risk and in clearing up any misconceptions about the nature of the illness itself and the help that is available. A person with a serious mood disorder can still lead a rich and productive life, even one of high achievement. Furthermore, the availability of lithium, antidepressants, and psychological treatments has greatly reduced the frequency of attacks as well as the intensity of the highs and lows that were once an inescapable feature of this illness. As we learn more about the genetic bases of manic-depression, these treatments should improve, and preventative measures may also be discovered.

What about children in affected families? Should they be informed of their greater vulnerability? At what age? Should steps be taken to protect them? If so, what steps? The answers to these questions can be answered only on a case-by-case basis, and again, professional consultation can help.

AS MOOD GENES are identified, professional assistance may also include DNA testing to determine whether a person carries alleles that increase susceptibility to manic-depression. But such risk assessment will never reach the level of certainty that is possible for disorders such as Huntington's disease, in which the presence of an abnormal allele of a single causative gene tells the whole story. In the case of manic-depression, increased vulnerability will probably be correlated with a particular combination of mood-gene alleles instead of just a single one. Furthermore, a number of different alleles of the same mood gene may each contribute to increased vulnerability, just as hundreds of different alleles of the G6PD may each confer vulnerability to favism. Evaluation of a person's "profile" of mood-gene alleles may therefore require elaborate testing of a DNA sample.

Fortunately the detection of multiple alleles in an individual DNA sample can already be achieved in several ways. One promising method makes use of an array of *thousands* of different DNA probes—each "printed" (by a special chemical technique) at a specified spot on a tiny silicon chip (called, in a patented version, GeneChip) that is smaller than a postage stamp, each custom-made to detect a specific allele. To test for the presence or absence of a whole range of mood-gene alleles in a person's DNA, copies of bits of the relevant alleles would be made using the polymerase chain reaction (PCR). The mixture of bits of alleles would then be added to the chip and each would bind to the spot of probe designed to detect it. After the binding was completed, the mood-gene alleles that were present in the DNA sample would be detected by optically scanning each spot on the chip with a computer-controlled device. In this way, many different alleles of multiple mood genes could all be examined in a single DNA sample at a modest cost, to generate an individual "mood-gene profile."

But even an elaborate mood-gene profile won't make it possible to predict with certainty whether or not a particular person will develop a mood disorder. We know this because, as we have already seen, identical twins—who, being genetically identical, must both have exactly the same mood-gene profile—don't always both develop the same mood disorder. Nevertheless, if one twin has bipolar disorder, the identical twin has about 6 chances in 10 of developing the same disorder. Thus DNA testing may in principle make possible risk estimates as high as about 6 out of 10 chances of developing this form of manic-depression if the person's mood-gene profile has the necessary combination of alleles. DNA testing may also reveal intermediate levels of risk in people who have inherited only some of the relevant mood-gene alleles from an affected parent.

Prenatal DNA testing for mood genes—when it becomes practicable—could even raise the ethically controversial possibility of the abortion of fetuses deemed to be at risk for manic-depressive illness. With other diseases such as sickle cell anemia, for which risk to the fetus can already be predicted with great accuracy on the basis of DNA testing, abortion has become an established option. Furthermore, according to a survey of members of a support group for patients with manic-depression, their relatives, partners, and friends, more than half say they would want a prenatal test and would terminate a pregnancy if they knew without a doubt that the fetus would inherit a severe form of manic-depressive illness. Some members of this support group are so distressed by the thought of having to confront this dreaded disorder in their children that they say that they would not carry a fetus to term even if the child born were fated to have only a mild form of the disease. Though the experience with Huntington's disease tells us that people who say they intend to use genetic testing frequently don't go through with it, some prenatal mood-gene testing is bound to take place, and some fetuses found to be at high risk for manic-depression are bound to be aborted.

Should you find this shocking, you may be amazed to learn that there are parents who already terminate pregnancies because of a prenatally diagnosed trait *that has nothing at all to do with illness*—the sex of the fetus. And judging from their responses to a questionnaire about ethical dilemmas that was circulated in the late 1980s, a substantial proportion of medical geneticists explicitly condones this practice. Among the hypothetical cases they were asked to consider in this questionnaire was the following:

> A couple requests prenatal diagnosis for purposes of
> selecting the sex of the child. They already have four girls
> and are desperate for a boy. They say that if the fetus is a
> girl, they will abort it and will keep trying until they
> conceive a boy. They tell you that if you refuse to do
> prenatal diagnosis for sex selection, they will abort the fetus
> rather than risk having another girl. The clinic for which
> you work has *no* regulations prohibiting use of prenatal
> diagnosis for sex selection. What would you do?

When faced with this choice 62 percent of the 295 respondents from the United States said that they would either comply with the parents' request for prenatal testing or refer them to someone else who would. This is not to say they would all approve. Dorothy Wertz and her colleagues, who conducted this extensive survey, personally balked at the idea of using prenatal diagnosis for sex selection. In their view, "sex selection discredits the public image of prenatal diagnosis and of medical genetics, lends support to the campaigns of the anti-abortionists, and sets a precedent for parental choices on 'cosmetic' grounds." However, considering that many respondents believe that parents have the unconditional right "to determine the number, spacing and quality of their children" (according to Wertz's summary of the results), there seems little doubt then that they would be willing to provide parents with a test of fetal mood genes—no matter what use it might be put to.

The widespread opinion that people have an unconditional right to reproductive choice has led to a revival of interest in eugenics (from the Greek, "good in birth"), a term coined in 1883 by Francis Galton, whose early studies of "eminence" and other complex traits we have already considered. Believing that much of human diversity has a hereditary basis, Galton and his followers advocated the preferential breeding of the best and the brightest. But the many horrible consequences of the eugenics movement, which culminated in the involuntary sterilization of thousands of residents of mental institutions in the United States and the millions of murders by the Nazis, led to a widespread condemnation of this form of eugenics in the aftermath of World War II. With the advent of genetic testing, a new form of eugenics is arising that is based on individual reproductive choice instead of state-mandated coercion, and on an informed view of the likelihood of various outcomes in a particular fetus instead of a distorted view based on racial and social prejudices.

Among those who argue that we must prepare ourselves for a new eugenics based on genetic testing and personal reproductive choice is the philosopher Philip Kitcher. In his recent book, *The Lives to Come,* Kitcher predicts that we are headed for what he calls "utopian eugenics," which he traces back to British social reformers such as George Bernard Shaw. In this form of eugenics there would be no established policy about the nature of desirable offspring; but prenatal testing and reliable genetic information would be available to all, and people would be free to make individual decisions about reproduction based on their personal assessments of the consequences for their offspring and for society. As Kitcher points out, "once we have left the garden of genetic innocence, some form of eugenics is inescapable. . . ." James Watson, co-discoverer of the structure of DNA and a founder of the Human Genome Project, puts it this way: "If we could honestly promise young couples that we knew how to give them offspring with superior character, why should we assume they would decline?"

NEVERTHELESS, THE MAIN aim of finding mood genes is not eugenic: the main aim is to determine how they contribute to mood disorders by influencing aspects of brain function. From this knowledge will come better ways of subdividing this heterogeneous group of conditions, opening the possibility of tailor-made treatments for different forms. Finding mood genes will also set the stage for the development of new therapeutic drugs that alleviate symptoms and prevent attacks so effectively that any inclination to abortion of vulnerable fetuses may eventually disappear.

We need these new drugs because those we now have, though a godsend for many people, have significant drawbacks. For one thing, they may have uncomfortable side effects, from the tremors and memory loss found with lithium to the bouncing blood pressure or sexual impairment found with certain antidepressants. For another, they aren't always effective. Considering that not only lithium, but also the prototypes of the drugs we use to treat depression, were all discovered by accident, it is amazing that they work as well as they do.

The accidental discovery of two categories of antidepressants came just a few years after Cade stumbled upon the antimanic effect of lithium (in the curious manner that I already described). One category of antidepressants is descended from iproniazid, a drug that was developed to kill the microorganism that causes tuberculosis, but that happened also to relieve the depressed mood of some patients who received it. The larger and more widely used category is descended from imipramine, a drug that had been designed as an antihistamine, and was then unexpectedly found to alleviate major depression. From the time of their introduction in the mid-1950s, these drugs have revolutionized the treatment of depression. But in neither case was the initial discovery guided by even an inkling of how the drug might work.

Only later did it become apparent that iproniazid and imipramine have something in common: both prolong the actions of three amine neurotransmitters—norepinephrine, serotonin, and dopamine—that are released by nerve cells in brain circuits that control our emotions. Iproniazid does this by inhibiting the activity of an enzyme called monoamine oxidase (MAO), which terminates the actions of these amines by destroying them. Since the destruction is slowed down by the iproniazid, the amines continue to stimulate nerve cells for a longer period of time. There is another pathway by which amine neurotransmitters are inactivated, and that is by pumping them back into the nerve cells that released them, a process called reuptake. Imipramine's effect, similar to that of iproniazid, results from blocking reuptake of the amines. In this way, imipramine (and its descendants) keep the released amines active longer, allowing more sustained stimulation.

Once the shared effect of iproniazid and imipramine on the amine neurotransmitters was established, pharmaceutical companies scrambled to discover more selective variants of these drugs in the hope of finding some with better properties. The result was the development of a new class of amine reuptake blockers that prolong the action of only a single amine, serotonin, and are therefore called selective serotonin reuptake inhibitors, or SSRIs. Lacking many side effects of the earlier drugs, several chemically distinct SSRIs—marketed under trade names such as Prozac, Zoloft, and Paxil—now enjoy immense popularity as antidepressants (though they too have side effects, mainly on sexual function).

Even though enough has been learned about imipramine and iproniazid to guide these important refinements, we still don't really understand how any of these drugs actually clear up depression. One major puzzle is that drugs that selectively prolong the actions of either serotonin (that is, SSRIs) or norepinephrine (for example, desipramine) are equally useful antidepressants, although these two amine neurotransmitters have very different functions in the brain. Also very puzzling is the lag of several weeks before any of the antidepressants actually takes hold, even though each of their direct effects on amines in the brain is established immediately. And, to top things off, there is a similar lag of several weeks before lithium effectively blocks mania. Together these findings suggest that the therapeutic actions of all these drugs result from unidentified adaptive changes in the brain that take weeks to develop, and then somehow counteract the fundamental causes of the mood disorder. If that sounds to you like an acknowledgment of limited understanding, it surely is.

Which explains a major reason medical geneticists are so eager to identify mood genes: to use that knowledge to figure out exactly how these genes are related to neurotransmitters and their actions, thereby guiding improvements of the available drugs. In addition, knowing the identity and function of mood genes will provide the opportunity to develop whole new categories of drugs with completely different molecular targets. Such a change of direction is sorely needed. Aside from the adoption of a few drugs that were developed for epilepsy as treatments for mania (for example, carbamazepine, marketed as Tegretol, and sodium valproate, marketed as Depakote), there has been no new class of drugs for the treatment of mood disorders for more than forty years.

This is not to say that the path from mood-gene discovery to new drug treatments is going to be easy. Identification of genes involved in Alzheimer's disease has so far generated only some new ideas but no new treatments. Manic-depression may prove to be even more challenging, because interactions of multiple mood genes may be involved in the development of symptoms. Yet a drug that normalized the function of just one of them might be sufficient to tip the balance.

Though drug development is the main goal that drives the hunt for mood genes, DNA tests for vulnerability to manic-depression will eventually have an important place as well. Just as the *APOE* test is already used in the diagnosis of people with dementia, mood-gene tests are bound to be used for diagnostic purposes in people with mood disturbances. Such tests may prove to be of particular value in distinguishing subtypes of mood disorders that may respond to different treatments.

It is likely that mood-gene testing will at first be restricted to people who have had a manic or depressive episode, but will eventually be extended to asymptomatic relatives of people with mood disorders as well. And though there will be opposition—for the same reasons that *APOE* testing of asymptomatic people is presently opposed—it will surely wane when preventative measures are developed. In the long run a major benefit of mood-gene discovery may be the prevention of all symptoms of manic-depression—even initial attacks.

Which brings us back to Michael and his family.

11
"BETTER THAN WELL": SOME TROUBLE WITH MICHAEL

How fleeting are the wishes and efforts of man! how short his time! and consequently how poor will his products be, compared with those accumulated by nature during whole geological periods. Can we wonder, then, that nature's productions should be far "truer" in character than man's productions; that they should be infinitely better adapted to the most complex conditions of life, and should plainly bear the stamp of far higher workmanship?

—Charles Darwin (1859)

Several years ago, and more than a decade after his son Jerry's manic episode, Michael decided to go to his family doctor for a checkup. Mainly interested in a routine examination, he was also concerned that he wasn't feeling as spunky as he used to. Though still in his fifties, Michael felt that his productivity had begun to slip and that he seemed to be having less fun. Could there be something wrong?

Nothing alarming turned up at the checkup, but Michael wasn't satisfied. He wondered if there was a treatment for his waning zest for living. Wasn't there a remedy his doctor could provide?

What Michael had in mind was Prozac or one of the other new selective serotonin reuptake inhibitors (SSRIs) that were being widely discussed. Although he knew that he wasn't suffering from a serious depression, Michael had heard so much about the favorable effects of the SSRIs that he decided to ask for a prescription. His main impetus was the hope that an antidepressant might elevate him to a new plane, making him (in the phrase of a patient described in Peter Kramer's widely read book *Listening to Prozac*) "better than well." Even Marcia, his usually cautious wife, had no objection to his making this personal pharmacological experiment, since Michael did seem particularly moody.

Michael's doctor agreed to give it a try. Although Michael's discontent was not an officially approved indication for administering antidepressants, thousands of physicians had by the mid-1990s become accustomed to prescribing them in the hope of alleviating complaints of unhappiness. Having known Michael for many years and keenly aware of his growing moodiness, his doctor was, in fact, frankly curious to see if the drug would have any effect. He decided to prescribe a relatively small dose, half that normally used in the treatment of major depression.

SIX WEEKS LATER I received an urgent call from Marcia. She was having some trouble with Michael.

She had begun to worry after Michael had taken the first few daily doses of the drug, which seemed to make him unusually restless. But she didn't become really concerned until he had been taking it for about a month, when major changes became apparent. The first one she noticed was that Michael, normally a sound sleeper, was up most of the night. Then she learned that he had been having shouting arguments with several colleagues at work and that he had been liquidating his retirement fund in order to raise cash for a very speculative investment. When a neighbor mentioned that she had seen Michael having dinner at a restaurant with a young woman who lived down the street, Marcia called to ask if I might be able to help.

My visit to Michael's office the following day provided shocking confirmation for what I had already suspected. Michael —whom I had always known as relaxed and easygoing—was obviously manic. Bombastic in style and grandiose in his plans for the future, he had in a matter of a few weeks embarked on a new life course that would almost certainly lead him to personal and financial ruin. In fact, if he didn't calm down at once, the damage might be irreversible.

Because it appeared so likely that the immediate cause of his manic behavior was the antidepressant drug, I (in concert with his doctor and with the fierce insistence of Marcia) persuaded him to discontinue it. Not without a fight: with the pills Michael felt himself—in his words—"*more* than better than well." But his supply was about to run out, and no more would be coming. So he reluctantly agreed to let Marcia flush the remaining pills down the toilet, and to see a psychiatrist once a week until he settled down. In about a month, he managed to restore both his marriage and his retirement fund, and he was back to his usual self.

Michael was not the first person with an apparent genetic vulnerability to manic-depression who had been tipped into mania by an antidepressant drug. Hagop Akiskal, a psychiatrist at the University of California, San Diego, regards drug-induced mania as a characteristic feature of an increasingly common form of manic-depressive illness that he calls "bipolar III." But unlike most others who have been affected in this way, and who afterward are simply upset about the drug-induced disruption of their lives, Michael was also intrigued by what had happened.

Before this turn of events, Michael already considered it extremely likely that he was the source of the mood-gene alleles that had led his son, Jerry, to manic-depression. But he also recognized that there was still a chance that Marcia was really the major (or sole) contributor of the alleles that were critical in Jerry's case, even though neither she nor any known sibling or ancestor had been diagnosed with a mood disorder. Now that Michael had direct evidence of his own propensity to mania, it was hard to resist the conclusion that he had indeed inherited at least some of the relevant mood-gene alleles from Flora and passed them on to Jerry.

It was a conclusion that he subsequently formalized by sending me an updated version of the pedigree that he had first constructed almost fifteen years before. Now, in place of the open square that before had represented Michael, was a square that was heavily blackened. As far as Michael was concerned, the drug-induced mania that he had experienced was phenotype enough to warrant more than a mere question mark. And while he deeply regretted the pain that he had brought to Marcia by his misconduct while taking the antidepressant, Michael was also fascinated by the vulnerability it uncovered: he felt that it conclusively demonstrated the mode of transmission of the mood disorder in his family.

THERE IS, OF COURSE, great irony in Michael's drug-induced mania. Irony in the fact that he, while advocating the newest genetic technology for the study of mood disorders, fell victim to the fruits of the next-to-newest technology—an SSRI antidepressant. Irony in the fact that he, wanting to use this technology to become "better than well," came close to wrecking his life.

But instead of simply being viewed as an example of the dangers inherent in all technologies, Michael also can be viewed as a victim of an overarching technology that is still in an early stage of development. For had the hunt for mood genes already succeeded, Michael could have been tested before receiving a prescription for the drug; and assuming that he had been found to be genetically vulnerable to mania, this particular drug could have been replaced with other means of dealing with his complaint of discontent, based in part on a knowledge of his particular genotype.

Someday such knowledge may come not only through studies of manic-depression but also from other research directly concerned with factors that influence a personal sense of what psychologists have formally defined as "well being"— exactly the sense that led Michael to the SSRI in the first place. Already there is evidence, mainly from comparisons of pairs of identical and fraternal twins—some raised together and others separated after birth—that genes play a major role in the sense of "well being" that Michael was hoping to influence with a drug. These studies also indicate that unlike manic-depression, which appears to be affected by the combined actions of alleles of only a few genes (perhaps as few as three), "well being" is affected by combinations of alleles of a much larger number of genes—combinations that must, of course, be shared by identical twins (accounting in part for their remarkably similar "well being") but that are very unlikely to be shared by fraternal twins (accounting for the fact that their "well being" is no more similar than that of two unrelated people). Nevertheless, despite this genetic complexity, some have argued that the identification of alleles that have a measurable impact on a sense of "well being" may be achieved in the not too distant future, an undertaking that would be greatly facilitated by using the chips and other procedures that are presently being developed to simultaneously measure thousands of alleles in a single DNA sample.

In the end it would not be surprising if some of the genes that contribute to our sense of "well being" also influence the vulnerability to manic-depression. It has even been argued that a study of genetic influences on temperament may prove to be as valuable as linkage studies in the identification of alleles that contribute to mental disorders. Conversely, the hunt for mood genes, driven by the pressing need to cope with an illness that leads many people to disaster, is also likely to bring ways of dealing with our normal emotions a bit closer to our grasp.

This activist view of behavioral genetic research is far different from the fatalistic one that prevailed before genetics could be studied at the level of specific variations in DNA. Not many years ago, saying that a behavioral tendency was genetic was generally taken to mean that it was permanently ingrained and unchangeable. Now we know that finding alleles that influence particular behavioral variations may not just be used to foretell our destinies—but also to forestall them.

NOTES

PROLOGUE

1   "It will be much, some day.": Charles Dickens, "A Curious Dance around a Curious Tree," *Household Words,* January 1852.

2   Among those affected . . . Charles Dickens and his father John: W. Russell Brain, a British neurologist, describes Charles Dickens's recurrent episodes of severe depression and "his prevailing mood of elation, . . . which at first was canalized into his prolific inventiveness, in later life squandered in his . . . parties, and amateur theatricals" (Brain, 1949). In a detailed review of Charles Dickens's mood fluctuations, D. J. Hershman and J. Lieb conclude that not only he but also his father, John, suffered from manic-depressive illness. As they describe John Dickens (whose flowery speech they consider a model for Mr. Micawber's in *David Copperfield*), he "had one fatal manic flaw that opened a chasm beneath his feet. He loved to spend money, not only his own but also everyone else's, including, eventually, that of Charles's publishers. John became a chronic bankrupt, and Charles's contributions kept him out of jail more than once" (Hershman and Lieb, 1988).

1. MICHAEL'S FAMILY

7   Michael's family: Details about Michael and his family have been altered to protect their anonymity.

9   five times as many people: Severe depression affects at least 7 percent of American women and 3 percent of American men at some time in their lives (Weissman et al., 1996). Milder forms of depression are even more common.

10   AKU is hereditary: The story of Garrod's pioneering work on alkaptonuria is described in Bearn and Miller (1979) and Bearn (1993).

Later developments are in Beighton (1993), Scriver et al. (1995), and Scriver (1996).

12 equivalent of alkapton: An example of an equivalent of an alkapton that may give rise to prominent mental symptoms is described in the discussion of porphyria in Chapter 4.

12 John Cade's . . . discovery: Cade recalls his discovery in Ayd and Blackwell (1970).

15 the genetic material: Watson and Crick (1953).

16 the Central Dogma: The story of Crick's formulation of the Central Dogma is described by Judson (1979).

18 Having been at the right place at the right time: Nirenberg et al. (1963).

2. A SINGLE MORBID PROCESS?

25 book on the hundred most influential scientists: Simmons (1996). The author is a biographer who consulted with a number of scientists in developing his ranking. The following scientists who are mentioned in this book made Simmons's top 100: Isaac Newton, 1; Charles Darwin, 4; Sigmund Freud, 6; Linus Pauling, 16; Rudolf Virchow, 17; Francis Crick, 33; Carl Gauss, 41; James Watson, 49; Gregor Mendel, 60; Thomas Hunt Morgan, 62; Paul Ehrlich, 64; Ernst Mayr, 65; Frederick Sanger, 72; Lucretius, 73; Emil Kraepelin, 92; Francis Galton, 94; Wilhelm Wundt, 99.

26 "somatic affection": Freud (1917).

26 Emil Kraepelin: Kraepelin's life and work is described by Alexander and Selesnick (1966), Shorter (1996), and Zilboorg (1941). His work on classification is reviewed by Berrios and Hauser (1988).

28 how this degeneration comes about: Selkoe (1997).

29 "single morbid process": Kraepelin (1915).

30 "already fallen ill": Ibid.

32 horrors that followed: Lifton (1986), Muller-Hill (1988), and Gottesman and Bertelsen (1996).

32 research with . . . Ernst Brücke: Freud's metamorphosis from biologist to psychoanalyst is described by Sulloway (1979).

32 "I think I was happier": Letter to Karl Abraham, September 21, 1924, referring to work Freud published in 1878, cited in Sulloway (1979).

32 Project for a Scientific Psychology: Freud (1895).

35 "melancholia to mania": Freud (1917).

36   "all its desires": Freud (1933).

36   "names . . . be discarded": Menninger (1963).

37   "a rigid and sterile codifier": Alexander and Selesnick (1966).

37   "problems in living": Szasz (1961).

39   "mischievous and interfering": Cade in Ayd and Blackwell (1970).

39   lithium . . . sometimes had distressing side effects: Goodwin and Jamison (1990) summarize the most frequently reported subjective side effects of lithium (with percent of patients affected): excessive thirst (36 percent), excessive urination (30 percent), memory problems (28 percent), tremor (27 percent), and weight gain (19 percent). With prolonged treatment there may be reduced functioning of the thyroid gland and the kidney.

39   don't really know how it works: Manji et al. (1995).

39   series of useful drugs: Barondes (1993) and Snyder (1996).

40   agreed to pay $650,000: Karel (1994).

40   "not to the natural sciences": Stone (1997).

41   Schedule for Affective Disorders and Schizophrenia (SADS): Endicott and Spitzer (1978).

41   they campaigned actively: Kirk and Kutchins (1992) describe the history of the creation of *DSM-III* and *DSM-IV.*

42   "polarity": Leonhard (1957).

42   "four basic patterns of symptoms": American Psychiatric Association (1994).

43   make use of feelings: Nesse and Williams (1995).

45   summarized in 1986: McGuffin and Katz (1986). Similar results were found in a study of an additional 955 index cases by Andreasen et al. (1987). See also Blacker and Tsuang (1992) and Blacker et al. (1996).

46   0.5–0.9 percent of both men and women in the United States and Western Europe: Weissman et al. (1996). In this study of mood disorders in ten countries, the lifetime rates (or risks) of bipolar disorder, type I, ranged from 0.3 per hundred (in Taiwan) to 1.5 per hundred (in New Zealand). The usual time of onset was late adolescence or early adulthood, and in each country males and females were affected about equally.

46   7 percent in women, 3 percent in men: Ibid.

46   early-onset major depression: Weissman et al. (1984). See also Weissman et al. (1988) and Kovacs et al. (1997).

47   more than tenfold the lifetime risk: This point was first emphasized by a British pioneer in psychiatric genetics, Lionel Penrose (1953):
"When the difference between the familial and population incidence is
found to be large, say tenfold or more, the result is so striking that the
probability of genetic causation appears to be high even in data compiled
with little attention to detail." See also Risch (1990a,b).

47   bipolar disorder, type I has become the favorite target: This conclusion is strongly supported by the participants in the National Institute of
Mental Health Collaborative Program on the Psychobiology of Depression,
as reported by Winokur et al. (1995).

3. The Astonishing Leap from Traits to Genes

49   Gregor Mendel: Mendel's life and discoveries are described in
Sturtevant (1965).

51   "in the hybrids": Mendel (1866).

55   Some Offspring of Two of Mendel's Pea Plants: In the diagram
the parents and offspring are shown as either male (square) or female (cir-
cle) to facilitate the analogy with people, although each pea plant really has
both male and female sex organs which are experimentally manipulated to
permit interbreeding.

58   human giants or dwarfs: Examples of alleles that can be major
factors in producing giants, as in Marfan's syndrome (page 77), or dwarfs,
as in achondroplasia (page 60), are described in detail by Beighton (1993).

59   sugar into starch: Bhattacharyya et al. (1990, 1993).

60   achondroplasia: Beighton (1993).

60   pycnodysostosis: Maroteaux and Lamy (1965).

61   "had understood him": Eiseley (1958).

4. From Peas to People

64   a number of letters: Bearn and Miller (1979) and Bearn (1993).

64   "of such phenomena": Garrod (1902).

68   the gene responsible for AKU: Current information about this
gene is in Scriver et al. (1995) and Scriver (1996).

69   extending it to several other diseases: Garrod (1909).

69   generally called "causative": In medical genetics genes are called
causative if inheritance of a dominant allele (as in Huntington's disease) or
a pair of recessive alleles (as in AKU) is both necessary (setting aside issues
of genetic heterogeneity) and (pretty much) sufficient for the disease to
develop, assuming a "normal" (or "average") genetic and environmental

background and a long enough life span. As Mayr (1961) has pointed out, there are several ways of thinking about causality in living things, and there are several practical implications of finding a cause, including prediction and explanation. For example, if a person inherits particular alleles of a causative gene we can predict with considerable confidence that the person will eventually develop the disease; and by knowing the biological function of the normal allele of a causative gene we can explain the dysfunction that comes with inheritance of the abnormal one (or pair), and try to take steps to correct it.

70   "are seldom attacked": Huntington (1872).

72   "susceptibility" genes: Unlike alleles of causative genes (also called "genes of major effect"), alleles of susceptibility genes (also called "genes of modest effect") contribute to the development of disease only by acting in concert with critical alleles of other susceptibility genes as well as critical environmental factors. The blurry boundaries between causative and susceptibility genes will become apparent in the discussion of penetrance and variable expressivity later in this chapter.

72   aspects of environment: Some environmental factors that influence the onset of complex diseases are well known. For example, overeating increases the chances of developing diabetes, a complex disease also known to be influenced by multiple susceptibility genes. But other environmental factors have been very difficult to identify. Included in this category are random cellular events during embryonic development and throughout life.

73   probably porphyria: Macalpine and Hunter (1966). These authors later raised the possibility that George III might have had a different form of inherited porphyria called variegate porphyria.

73   "not have been psychotic": The citation in Macalpine and Hunter (1969) is from Manfred S. Guttmacher's *America's Last King: An Interpretation of the Madness of George III* (New York: Scribner's, 1941).

74   an abnormal one (I): The abnormal allele (I) is capitalized because, if penetrant, it is dominant.

75   a single dominant allelle . . . in the transmission of this disease: Spence et al. (1995). Were manic–depressive illness transmitted by a single dominant allele of a mood gene (for example, M), 50 percent of the children of an Mm parent and an mm parent would have the genotype Mm and be genetically vulnerable to the development of this mood disorder. Since studies of many affected families indicate that only 8 percent of the children of a parent with manic–depressive illness (narrowly defined as bipolar disorder, type I) develop this mood disorder, this could be taken to mean that the penetrance of M is 16 percent. Though low, this penetrance

exceeds that of the 10 percent penetrance of the *I* allele that gives rise to porphyria.

76   at least three mood genes must work together: Craddock et al. (1995). One reason to invoke the joint action of alleles of at least three mood genes (for example, *A*, *B*, and *C*) is to help explain the observation that only about 8 percent of the children of a parent with narrowly defined manic-depressive illness (that is, bipolar disorder, type I) develop this disorder. For example, were the genotype of a parent with manic-depression *Aa*, *Bb*, *Cc*, and the genotype of the other (unaffected) parent *aa*, *bb*, *cc*, only 12.5 percent of their children would have the genotype *Aa*, *Bb*, *Cc*—the genotype that confers vulnerability to the mood disorder. This theoretical figure—12.5 percent—is not much greater than 8 percent—the actual percentage of children of an affected parent who are also affected. The difference between 12.5 percent and 8 percent is readily attributable to nonpenetrance. See also McGuffin et al. (1994) and Risch (1990a,b).

76   "by his feet": Cited in Beighton (1993).

77   protein called fibrillin: McKusick (1991) and Beighton (1993).

77   retrospective diagnosticians: Ibid.

78   various environmental factors: Ways of evaluating the relative contributions of susceptibility genes and environmental factors to the development of mental disorders are discussed by Plomin, Chipuer et al. (1994), Plomin, Owen et al. (1994), Kendler (1995), and Plomin et al. (1997).

78   might be a mild expression: Akiskal and Akiskal (1992).

78   Francis Galton: An account of Galton's life is given in Kevles (1986).

79   "Science of man": Galton (1889).

79   seemed inapplicable to the complex human traits: Froggat and Nevin (1971).

79   theoretical work of Ronald Fisher: Fisher (1918).

80   human personality: Galton (1865).

80   characteristic was familial: Galton (1869).

80   identical twins . . . raised apart: Bouchard and Propping (1993), Plomin, Chipuer et al. (1994), Plomin et al. (1997), and Bouchard (1994).

81   "The History of Twins": Galton (1876).

81   appears to be the case: Bouchard and Propping (1993), Plomin and Daniels (1987), and Plomin et al. (1997).

81   group led by Aksel Bertelsen: Bertelsen et al. (1977).

83   high blood pressure: Lifton (1996).

83   diabetes: Todd (1996).

5. SOME TOOLS FOR THE HUNT

85   "due to our ignorance": Cited by Shine and Wrobel (1976), p. vi.

85   "a force of nature": Judson (1979).

87   *direct reflections of the alleles:* Pauling et al. (1949).

88   single amino acid substitution: Ingram (1957).

89   gene that encodes the serotonin transporter: Ogilvie et al. (1996) and Kunugi et al. (1997) describe some of the conflicting evidence about the relationship between susceptibility to mood disorders and alleles of the gene that encodes the serotonin transporter protein.

89   tyrosine hydroxylase: Turecki et al. (1997).

90   functions . . . are still unknown: Many new proteins are being discovered on the basis of the DNA sequences that encode them in the course of the ongoing systematic study of human DNA, the Human Genome Project. In some cases it has been easy to figure out the biological functions of proteins that were discovered in this way, but in many other cases their functions remain unknown.

91   Thomas Hunt Morgan: For details about Morgan's life, see Shine and Wrobel (1976).

92   "mustered enough strength": Shine and Wrobel (1976).

93   normal dominant allele, *R*: Because genes are generally named for a mutant phenotype, the official name for this gene is *white* (with alleles *W* and *w*), rather than *red* (with alleles *R* and *r*). But readers of early drafts of this book who were unfamiliar with this convention found it easier to follow this complex explanation if the gene was called *red* with alleles *R* and *r*.

99   the first map of the X chromosome: Sturtevant (1913).

102   "a statistically significant finding": For a discussion of interpretations of such data, see Lander and Kruglyak (1995).

104   ability to roll the tongue was controlled by . . . a single gene: Sturtevant (1940).

104   modifiable by learning: Sturtevant (1965).

105   "an established Mendelian case": Ibid.

106   restriction enzymes: For a discussion of restriction enzymes, see Watson et al. (1992).

107   milestones in a genetic map: Kan and Dozy (1978).

108   a now classic paper: Botstein et al. (1980).

109   short tandem repeat polymorphisms: Weber and May (1989).

109   polymerase chain reaction: For a discussion of PCR, see Watson et al. (1992).

109   specific chromosomal loci: Cooperative Human Linkage Center et al. (1994) and Dib et al. (1996).

109   single nucleotide polymorphisms: The advantage of single nucleotide polymorphisms (snips) is that techniques have been developed to measure *simultaneously* tens of thousands of individual alleles in a single DNA sample, thereby making possible the rapid and inexpensive evaluation of thousands of loci in one fell swoop. This technology is presently being used to generate human genetic maps that will greatly simplify genetic screening (Kruglyak, 1997 and Collins et al., 1997).

110   skillful detective work: The zip codes identified by the linkage approach usually cover a region of at least a million base pairs, an area that contains, on the average, about twenty-five genes. Because we don't yet know the addresses or identities of most of the genes in a particular zip code, identifying a gene such as *M* now depends on a laborious examination of all the DNA in the region for evidence of a mutation. As the sequencing of the human genome proceeds (completion of this aspect of the Human Genome Project is expected by about 2005), we keep learning the addresses and identities of more and more genes, which is progressively simplifying the task of going from linkage to identification of the gene involved in a particular disease. This mass of knowledge about individual genes will eventually make possible still other approaches (Risch and Merikangas, 1996; Lander, 1996).

6. HUNTING WITHOUT A MAP

115   the founder . . . thought to be a European sailor: In her poignant memoir of her family's role in the hunt for the genetic basis of Huntington's disease, Alice Wexler describes her attempt to identify Antonio Justo Doria, a Spanish sailor who had arrived in Venezuela in the middle of the nineteenth century, as the source of the genetic abnormality in this population. Though Doria himself was ruled out as the source in this process, it remains likely that all affected people in the Lake Maracaibo region are descendants of a single ancestor. Wexler (1995).

117   odds that . . . linkage is real: Lander and Kruglyak (1995).

117   "chromosomal location of the gene defect": Gusella et al. (1983).

118   large team of scientists: Huntington's Disease Collaborative Research Group (1993).

118   a predictive test: In the absence of an effective treatment for Huntington's disease, most relatives of Huntington's disease patients have chosen not to be tested. Some of the issues raised by the availability of this test are summarized by Nancy Wexler in Kevles and Hood (1992). Evidence that both positive and negative test results may have psychological benefits is presented in the Canadian Collaborative Study of Predictive Testing (1992).

119   in the eighteenth century: McKusick (1978).

119   about fifty-fifty: McKusick (1991).

119   Galton-Garrod Society: Another founder, Barton Childs pointed out that "it was not until the early 60s that the number of papers [being published in medical journals] under such headings as human genetics, chromosomes and human heredity surpassed the number devoted to hernias." Childs (1973).

120   suicides . . . in only two families: Egeland and Sussex (1985).

123   a locus for manic-depressive illness: Egeland et al. (1987).

124   "of chromosome 11": Ibid.

124   failed to find linkage: Detera-Wadleigh et al. (1987) and Hodgkinson et al. (1987).

124   commentary on this body of work: Robertson (1987).

125   shattered by new data: Kelsoe et al. (1989).

126   "was in fact just chance": Robertson (1989).

126   seemingly compelling evidence: Baron et al. (1987).

128   no longer supported linkage to the X chromosome: Baron et al. (1993).

129   map . . . with thousands of closely spaced markers: Cooperative Human Linkage Center et al. (1994) and Dib et al. (1996).

129   alternative experimental and statistical methods: One popular alternative method—the sib-pair method—is designed to identify mood-gene-containing chromosome segments shared (more often than would be expected by chance) by pairs of siblings who both have manic-depression. The advantage of this method is that it is much easier to find pairs of affected siblings than to find families with large numbers of affected relatives. But it may be necessary to study hundreds or even thousands of pairs of affected siblings to demonstrate linkage. Groups of investigators are presently accumulating DNA samples from several thousand sib-pairs with manic-depression to hunt for mood genes. For a discussion of other alternative methods see Lander and Schork (1994), Lander and Kruglyak (1995), and Risch and Merikangas (1996).

130   a map based on snip markers: Kruglyak (1997) and Collins et al. (1997).

7. ANA'S FAMILY

131   "find the materials": Cited in Kevles (1986), p. 14.

136   DNA markers . . . failed to support the initial results: Baron et al. (1993).

136   eighty-six Spanish families: Escamilla et al. (1996).

140   deafness . . . traced back to a common ancestor: Leon et al. (1992).

141   Schedule for Affective Disorders and Schizophrenia (SADS): Endicott and Spitzer (1978).

141   Diagnostic Instrument for Genetic Studies (DIGS): Nurnberger et al. (1994).

142   Family 1: Freimer et al. (1996a).

144   Family 4: Ibid.

8. HOT SPOTS IN THE GENOME

145   "Whenever you can, count": Cited in Kevles (1986), p. 7.

146   six thousand markers on the map: Cooperative Human Linkage Center et al. (1994).

148   measurement of the lengths of the stirps: After millions of copies of these small polymorphic regions of DNA are made by the polymerase chain reaction (PCR), each sample is electrophoresed on a gel. For each stirp the alleles that each sample contains are revealed as bands of DNA. These are separated by electrophoresis on the basis of the number of CA repeats in the particular allele. The band that represents an allele with a small number of repeats, such as CACACA, migrates more rapidly than the band that represents an allele with a larger number of repeats, such as CACA-CACA. For details, see Weber and May (1989) and Watson et al. (1992).

148   met the screening criteria for . . . both families: McInnes et al. (1996).

148   shared by 23 of the 26 people with manic-depressive illness: Freimer et al. (1996b).

150   in a large Scottish family: Blackwood et al. (1996).

150   genome screen of Old Order Amish pedigree 110: Ginns et al. (1996).

151   Their commentary: Risch and Botstein (1996).

152   Berrettini. . . . found some evidence of linkage: Berrettini et al. (1994).

152   also examined chromosome 18: Stine et al. (1995).

152   a Belgian group: DeBruyn et al. (1996).

153   a Utah group: Coon et al. (1996).

154   the gene for a hereditary form of deafness: Leon et al. (1992) and Lynch et al. (1997). The causative gene, *DFNA1,* encodes a protein that is believed to influence the structure and sound-receiving properties of the hair cells of the inner ear. Neither the identity nor the function of this protein was suspected until a mutation in *DFNA1* was found by Lynch et al. by means of the linkage approach. *DFNA1* is only one of forty genes for inherited "nonsyndromal deafness" (inherited deafness without other observable abnormalities) whose locations have been mapped by linkage, and only the sixth case in which the gene has been found. The existence of at least forty different genes which, if mutated, can give rise to the same phenotype— deafness—is a particularly notable example of genetic heterogeneity.

155   the DNA of forty-eight of them: Escamilla et al. (unpublished result).

158   Alzheimer's disease . . . traceable to a single founder: Bird et al. (1989).

158   Alzheimer's disease attributable to different causative genes: Lendon et al. (1997).

159   hot spot on chromosome 19: Pericak-Vance et al. (1991).

161   "does not foretell disease": Post et al. (1997).

162   experts . . . are prepared to change their minds: Roses (1995, 1997).

162   blows to the head: Jordan et al. (1997). Other evidence for synergistic effects of traumatic head injury and *APOE4* is presented by Mayeux et al. (1995).

163   interactions of alleles of susceptibility genes: Recent studies have identified the locus of another susceptibility gene for Alzheimer's disease on chromosome 12 (Pericak-Vance et al., 1997).

9. BOTH GOOD SEED AND BAD

166   evolution of genetic variations: Weiss (1995) and Nesse and Williams (1995).

166   an advantage so great that it outweighed the lethal threat: Weatherall, D. J.; Clegg, J. B.; Higgs, D. R.; and Wood, W., in Scriver et al. (1995) pp. 3417–3484.

168   thalassemia: Ibid.

168   mutations in the gene that encodes G6PD: Luzzatto, L., and Mehta, A., in Scriver et al. (1995), pp. 3367–3398.

169   hawk–dove game: Maynard Smith (1989).

170   study of male territorial and sexual behavior: Sinervo and Lively (1996).

172   by alleles of one gene: Barry Sinervo, personal communication.

173   benefits that certain of their relatives enjoy: Andreasen (1987) and Richards et al. (1988).

173   to maintain a high frequency of these alleles: Weiss (1995) and Nesse and Williams (1995) discuss the mixture of advantages and disadvantages of certain alleles. Early papers on the view that alleles that increase susceptibility to mental disorders also may confer some benefits were published by Huxley et al. (1964) and Gardner (1982). Also see McGuire et al. (1992).

173   Newton . . . manic-depression: Though retrospective diagnoses are notoriously subject to error, Hershman and Lieb (1988) present a persuasive case that Newton (whose life events were well documented because of his early fame) suffered from manic-depressive illness.

174   "discomposure in head, mind, and both": Cited in Hershman and Lieb (1988).

174   The Wall Street Journal: Jenkins, Jr. (1996).

10. GRAPPLING WITH FATE

178   studies of large numbers of affected families: McGuffin and Katz (1986) and Weissman et al. (1984, 1996).

178   other summaries of family studies: Andreasen et al. (1987), Tsuang and Faraone (1990), E. S. Gershon in Goodwin and Jamison (1990), and Gershon et al. (1982).

180   GeneChip: GeneChip is manufactured by Affymetrix, Santa Clara, California. Alternative methods for simultaneous screening of large numbers of alleles are also being developed (Collins et al., 1997).

181   computer-controlled device: Editorial, Nature Genetics (1996) and Fodor (1997).

181   identical twins . . . don't always develop the same mood disorder: Bertelsen et al. (1977).

181   a survey of members of a support group: Smith et al. (1996).

181 say they intend . . . don't go through with it: Wexler (1995) and Canadian Collaborative Study of Predictive Testing (1992).

182 questionnaire about ethical dilemmas: Wertz et al. (1990).

183 involuntary sterilizations . . . in the United States: Kevles (1986).

183 "some form of eugenics is inescapable": Kitcher (1996).

183 "why should we assume they would decline?": James Watson, cited by Kevin Davies (1997). *Nature, 390* (6 November 1997), 33.

184 main aim . . . is not eugenic: Even in China, which enacted a Maternal and Infant Health Care Law in 1994 that seeks to prevent people with manic-depression from marrying or having children and that could, in principle, lead to mandatory testing for mood genes, it is unlikely that the government will really mount a sustained eugenic campaign of this type. Veronica Pearson of the University of Hong Kong, who has looked into this, reports: "I have never come across even one incidence of a mentally ill person being forbidden to marry and have children despite extensive experience of Chinese psychiatric hospitals. The existing regulations are largely ignored . . . the resources are not there to perform the necessary examinations. Privately, Chinese psychiatrists tell me they do not have the heart for such work." Pearson (1995).

184 uncomfortable side effects: Each of the drugs used to treat depression or mania has significant side effects. The main subjective side effects of lithium include excessive thirst and urination, memory difficulties, tremor, and weight gain. A major side effect of both the imipramine and iproniazid categories of antidepressants is light-headedness and fainting. People who take monoamine oxidase inhibitors must avoid many foods such as aged cheeses and red wine, which contain amines that can lead to attacks of high blood pressure. The many other side effects of these drugs are discussed in detail in Goodwin and Jamison (1990) and Schatzberg and Nemeroff (1995).

184 prolong the actions of three amine neurotransmitters: Snyder (1996) and Barondes (1993).

186 adaptive changes in the brain: Duman et al. (1997).

186 some new ideas but no new treatments: Selkoe (1997).

11. "Better than Well": Some Trouble with Michael

191 "bipolar III": Akiskal (1987).

192 psychologists have formally defined as "well being": Lykken and Tellegen (1996). "Well being" is an attribute of personality that is mea-

sured on a scale that scores data from a self-rating questionnaire, the Multi-dimensional Personality Questionnaire. See also Hamer (1996).

193 perhaps as few as three: Craddock et al. (1995) argue that as few as three loci (genes and their alleles) may control vulnerability to manic-depression. Spence et al. (1995) argue that just one locus may be playing a major role.

193 "well being" is no more similar than that of two unrelated people: Tellegen et al. (1988) and Lyken and Tellegen (1996). The main finding is that "well being" (as measured by responses to a questionnaire) is similar among pairs of identical twins, whether reared together or apart, but not in fraternal twins or other siblings. Lyken and Tellegen (1996) take their data to mean that "from 44% to 52% of the variance in Well-Being is associated with genetic variation," whereas "neither socioeconomic status, educational attainment, family income, marital status, nor an indication of religious commitment could account for more than about 3% of the variance in Well-Being." Lykken et al. (1992) interpret results of this type—that show great similarities in identical twins but no similarity in fraternal twins (despite the fact that fraternal twins, like other siblings, may be viewed as being "50% identical genetically")—as meaning that the trait reflects complex interactions of such a large number of alleles that it would be very unlikely for a pair of fraternal twins to inherit the same combination. This contrasts strikingly with the vulnerability to manic-depression, which presumably reflects the combined effects of a much smaller number of alleles which are often inherited in common by fraternal twins.

193 not too distant future: Hamer (1996).

193 simultaneously measure thousands of alleles: Chip technology, which was described briefly in Chapter 10, is explained in Editorial, *Nature Genetics* (1997) and Fodor (1997). References to alternative techniques for rapid examination of multiple alleles are provided in Collins et al. (1997).

193 as valuable as linkage studies: Cloninger et al. (1996).

SOURCES FOR FIGURES

31 Emil Kraepelin (1915). *Manic-Depressive Insanity and Paranoia*, translated 1921, by R. Mary Barclay and George M. Robertson. Edinburgh: E. and S. Livingstone.

47: Myrna M. Weissman et al. (1984). Onset of major depression in early adulthood. Increased familial loading and specificity. *Arch. Gen. Psychiatry 41*, 1136–1143.

142 and 144: Nelson B. Friemer et al. (1996a), An approach to investigating linkage for bipolar disorder using large Costa Rican pedi-

grees. *American Journal of Medical Genetics (Neuropsychiatric Genetics) 67*, 254–263.

146: Cooperative Human Linkage Center (CHLC) (1994). A comprehensive human linkage map with centimorgan density. *Science 265*, 2049–2073.

149: Alison L. McInnes et al. (1996). A complete genome screen for genes predisposing to severe bipolar disorder in two Costa Rican pedigrees. *Proc. Natl. Acad. Sci. USA 93*, 13060–13065.

161: Allen D. Roses (1996). Apolipoprotein E alleles as risk factors in Alzheimer's disease. *Ann. Rev. Med. 47*, 387–400.

BIBLIOGRAPHY

This bibliography has been confined to reviews, commentaries, and some primary papers that are cited in the text, as well as some important books. Included among the books are three histories of psychiatry (Alexander and Selesnick; Zilboorg; and a new one by Shorter); three histories of aspects of genetics (Sturtevant; Kevles; and Judson); two recent textbooks on human genetics (Strachan and Read; and Vogel and Motulsky) and three on other aspects of genetics (Griffiths et al.; Watson et al.; and Weiss); two detailed compendia of human genetic disorders (Beighton; and Scriver et al.); textbooks on behavorial genetics (Plomin et al.), psychiatric genetics (McGuffin et al.), and the genetics of mood disorders (Tsuang and Faraone); two overviews of mood disorders (Goodwin and Jamison; and Whybrow); and four books that deal with aspects of drug treatment of mental illness (Barondes; Bloom and Kupfer; Schatzberg and Nemeroff; and Snyder).

AKISKAL, HAGOP S. (1987). The milder spectrum of bipolar disorders: diagnostic, characterologic, and pharmacologic aspects. *Psychiat. Annals, 17,* 32–37.

AKISKAL, HAGOP S., and AKISKAL, KAREEN. (1992). Cyclothymic, hyperthymic, and depressive temperaments as subaffective variants of mood disorders. In A. Tasman, ed., *American Psychiatric Press Review of Psychiatry, 11,* 43–62. Washington, D.C.: American Psychiatric Press.

ALEXANDER, FRANZ G., and SELESNICK, SHELDON T. (1966). *The History of Psychiatry.* New York: Harper and Row.

AMERICAN PSYCHIATRIC ASSOCIATION. (1994). *Diagnostic and Statistical Manual of Mental Disorders* (4th ed.) (*DSM-IV*). Washington, D.C.: American Psychiatric Press.

ANDREASEN, N. C. (1987). Creativity and mental illness: prevalence rates in writers and their first-degree relatives. *Am. J. Psychiat., 144,* 1288–1292.

211

ANDREASEN, N. C.; RICE, J.; ENDICOTT, J.; CORYELL, W.; GROVE, W. M.; and REICH, T. (1987). Familial rates of affective disorder: a report from the National Institute of Mental Health Collaborative Study. *Arch. Gen. Psychiat., 44,* 461–469.

AYD, FRANK J., and BLACKWELL, BARRY, eds. (1970). *Discoveries in Biological Psychiatry.* Philadelphia: J. B. Lippincott.

BARON, MIRON; FREIMER, NELSON B.; RISCH, NEIL; LERER, BERNARD; ALEXANDER, JOYCE R.; STRAUB, RICHARD E.; ASOKAN, SUSHA; DAS, KAMNA; PETERSON, AMY; AMOS, JEAN; ENDICOTT, JEAN; OTT, JURG; and GILLIAM, CONRAD. (1993). Diminished support for linkage between manic depressive illness and X-chromosome markers in three Israeli pedigrees. *Nature Genet., 3,* 49–55.

BARON, MIRON; RISCH, NEIL; HAMBURGER, RAHEL; MANDEL, BATSHEVA; KUSHNER, STUART; NEWMAN, MICHAEL; DRUMER, DOV; and BELMAKER, ROBERT H. (1987). Genetic linkage between X-chromosome markers and bipolar affective illness. *Nature, 326,* 289–292.

BARONDES, SAMUEL H. (1993). *Molecules and Mental Illness.* New York: Scientific American Library.

BEARN, ALEXANDER G. (1993). *Archibald Garrod and the Individuality of Man.* Oxford: Oxford University Press.

BEARN, A. G., and MILLER, E. D. (1979). Archibald Garrod and the development of the concept of inborn errors of metabolism. *Bulletin of the History of Medicine, 53,* 315–328.

BEIGHTON, PETER, ed. (1993). *McKusick's Hereditable Disorders of Connective Tissue* (5th ed.). St. Louis: Mosby.

BERRETTINI, WADE H.; FERRARO, THOMAS N.; GOLDIN, LYNN R.; WEEKS, DANIEL E.; DETERA-WADLEIGH, SEVILLA; NURNBERGER, JOHN I.; and GERSHON, ELLIOT S. (1994). Chromosome 18 markers and manic-depressive illness: evidence for a susceptibility gene. *Proc. Natl. Acad. Sci. USA, 91,* 5918–5921.

BERRIOS, G. E., and HAUSER, R. (1988). The early development of Kraepelin's ideas on classification: a conceptual history. *Psychological Med., 18,* 813–821.

BERTELSEN, A.; HARVALD, B.; and HAUGE, M. (1977). A Danish twin study of manic-depressive disorders. *Br. J. Psychiat., 130,* 330–351.

BHATTACHARYYA, MADAN; MARTIN, CATHIE; and SMITH, ALISON. (1993). The importance of starch biosynthesis in the wrinkled seed shape character of peas studied by Mendel. *Plant Molecular Biol., 22,* 525–531.

BHATTACHARYYA, MADAN K.; SMITH, ALISON M.; ELLIS, T. H. NOEL; HEDLEY, CLIFF; and MARTIN, CATHIE. (1990). The wrinkled-seed character of pea described by Mendel is caused by a transposon-like insertion in a gene encoding starch-branching enzyme. *Cell, 60,* 115–122.

BIRD, THOMAS D., and BENNETT, ROBIN L. (1995). Why do DNA testing? Practical and ethical implications of neurogenetic tests. *Annals of Neurology, 38,* 141–146.

BIRD, T. D.; LAMPE, T. H.; NEMENS, E. K.; MINER, G. W.; SUME, S. M.; and SCHELLENBERG, G. D. (1989). Familial Alzheimer's disease in American descendants of the Volga Germans: probable genetic founder effect. *Annals of Neurology, 25,* 12–25.

BLACKER, DEBORAH, and TSUANG, MING T. (1992). Contested boundaries of bipolar disorder and the limits of categorical diagnosis in psychiatry. *Am. J. Psychiat., 149,* 1473–1483.

BLACKER, DEBORAH; FARAONE, STEPHEN V.; ROSEN, AMY E.; GUROFF, JULIET J.; ADAMS, PHILIP; WEISSMAN, MYRNA M.; and GERSHON, ELLIOT S. (1996). Unipolar relatives in bipolar pedigrees: a search for elusive indicators of underlying bipolarity. *Am. J. Med. Genet. (Neuropsychiatric Genet.), 67,* 445–454.

BLACKWOOD, DOUGLAS H. R.; HE, LIN; MORRIS, STEWART W.; MCLEAN, ALAN; WHITTON, CLAIRE; THOMSON, MARIAN; WALKER, MAURA T.; WOODBURN, KIRSTIE; SHARP, CLIFF M.; WRIGHT, ALLAN F.; SHIBASAKI, YOSHIRO; ST. CLAIR, DAVID M.; PORTEOUS, DAVID J.; and MUIR, WALTER J. (1996). A locus for bipolar affective disorder on chromosome 4p. *Nature Genet., 12,* 427–430.

BLOOM, FLOYD E., and KUPFER, DAVID J., eds. (1995). *Psychopharmacology: The Fourth Generation of Progress.* New York: Raven Press.

BOTSTEIN, DAVID; WHITE, RAYMOND L.; SKOLNICK, MARK; and DAVIS, RONALD W. (1980). Construction of a genetic linkage map in man using restriction fragment length polymorphisms. *Am. J. Hum. Genet., 32,* 314–331.

BOUCHARD, JR., THOMAS J. (1994). Genes, environment, and personality. *Science, 264,* 1700–1701.

BOUCHARD, T. J., and PROPPING, P. eds. (1993). *Twins as a Tool of Behavioral Genetics.* Chichester, England: John Wiley.

BRAIN, W. RUSSELL. (1949). Authors and psychopaths. *Br. Med. J., 7,* 1427–1432.

BURTON, ROBERT. *The Anatomy of Melancholy.* (1621). Abridged version (1979), JOAN K. PETERS, ed. New York: Frederick Ungar Publishing Co.

CANADIAN COLLABORATIVE STUDY OF PREDICTIVE TESTING. (1992). The psychological consequences of predictive testing for Huntington's disease. *N. Engl. J. Med., 327,* 1401–1405.

CHILDS, BARTON. (1973). Garrod, Galton, and clinical medicine. *Yale J. Biol. Med., 46,* 297–313.

CLONINGER, C. ROBERT; ADOLFSSON, ROLK; and SVRAKIC, NENAND M. (1996). Mapping genes for human personality. *Nature Genet., 12,* 3–4.

COLLINS, FRANCIS S.; GUYER, MARK S.; CHAKRAVARTI, ARAVINDA. (1997). Variations on a theme: cataloging human DNA sequence variation. *Science, 278,* 1580–1581.

COON, HILARY; HOFF, M.; HOLIK, J.; HADLEY, D.; FANG, N.; REIMHERR, F.; WENDER, P.; and BYERLY, WILLIAM. (1996). Analysis of chromosome 18 DNA markers in multiplex pedigrees with manic depression. *Biol. Psychiat., 39,* 689–696.

COOPERATIVE HUMAN LINKAGE CENTER (CHLC): MURRAY, JEFFREY C.; BUETOW, KENNETH H.; WEBER, JAMES L.; LUDWIGSEN, SUSAN; SCHERPBIER HEDDEMA, TITIA; MANION, FRANK; QUILLEN, JOHN; SHEFFIELD, VAL C.; SUNDEN, SARA; AND DUYK, GEOFFREY M. GÉNÉTHON: WEISSENBACH, JEAN; GYAPAY, GABOR; DIB, COLETTE; MORISSETTE, JEAN; LATHROP, G. MARK; AND VIGNAL, ALAIN. UNIVERSITY OF UTAH: WHITE, RAY; MATSUNAMI, NORISADA; GERKEN, STEVEN; MELIS, ROBERTA; ALBERTSEN, HAND; PLAETKE, ROSEMARIE; AND ODELBERG, SHANNON. YALE UNIVERSITY: WARD, DAVID. CENTRE D'ETUDE DU POLYMORPHISME HUMAIN (CEPH): DAUSSET, JEAN; COHEN, DANIEL; AND CANN, HOWARD. (1994). A comprehensive human linkage map with centimorgan density. *Science, 265,* 2049–2073.

CORYELL, WILLIAM; ENDICOTT, JEAN; KELLER, MARTIN; ANDREASEN, NANCY; GROVE, WILLIAM; HIRSCHFELD, ROBERT M. A.; and SCHEFTNER, WILLIAM. (1989). Bipolar affective disorder and high achievement: a familial association. *Am. J. Psychiat., 146,* 983–989.

CRADDOCK, NICK; KHODEL, VLADIMIR; VAN EERDEWEGH, PAUL; and REICH, THEODORE. (1995). Mathematical limits of multilocus models: the genetic transmission of bipolar disorder. *Am. J. Hum. Genet., 57,* 690–702.

CRICK, FRANCIS. (1988). *What Mad Pursuit: A Personal View of Scientific Discovery.* New York: Basic Books.

DARWIN, CHARLES. (1859). *The Origin of Species.* (Repr. 1996). New York: Gramercy Books.

DE BRUYN, AN; SOUERY, DANIEL; MENDELBAUM, KARINE; MENDLEWICZ, JULIEN; and VAN BROECKHOVEN, CHRISTINE. (1996). Linkage analysis of families with bipolar illness and chromosome 18 markers. *Biol. Psychiat., 39,* 679–688.

DETERA-WADLEIGH, SEVILLA D.; BERRETTINI, WADE H.; GOLDIN, LYNN R.; BOORMAN, DENISE; ANDERSON, STEWART; GERSHON, ELLIOT S. (1987). Close linkage of c–Harvey-*ras*-1 and the insulin gene to affective disorder is ruled out in three North American pedigrees. *Nature, 325,* 806–808.

DIB, COLETTE; FAURE, SABINE; FIZAMES, CECILE; SAMSON, DELPHINE; DROUOT, NATHALIE; VIGNAL, ALAIN; MILLASSEU, PHILLIPPE; MARC, SOPHIE; HAZAN, JAMILE; SEBOUN, ERIC; LATHROP, MARK; GYAPAY, GABOR; MORISSETTE, JEAN, and WEISSENBACH, JEAN. (1996). A comprehensive genetic map of the human genome based on 5,264 microsatellites. *Nature, 380,* 152–154.

DUMAN, RONALD S.; HENINGER, GEORGE R.; NESTLER, ERIC J. (1997). A molecular and cellular theory of depression. *Arch. Gen. Psychiat., 54,* 597–607.

EDITORIAL. (1997). Brave new now. *Nature Genet., 15,* 1–2.

EDITORIAL. (1996). To affinity...and beyond! *Nature Genet., 14,* 367–370.

EGELAND, JANICE A.; GERHARD, DANIELA S.; PAULS, DAVID L.; SUSSEX, JAMES N.; KIDD, KENNETH K.; ALLEN, CLEONA R.; HOSTETTER, ABRAM M.; and HOUSMAN, DAVID E. (1987). Bipolar affective disorder linked to DNA markers on chromosome 11. *Nature, 325,* 783–787.

EGELAND, JANICE A., and SUSSEX, JAMES N. (1985). Suicide and family loading for affective disorders. *JAMA, 254,* 915–918.

EISELEY, LOREN. (1958). *Darwin's Century.* Garden City, N.Y.: Doubleday and Company.

ENDICOTT, JEAN, and SPITZER, ROBERT L. (1978). A diagnostic interview: the schedule for affective disorders and schizophrenia. *Arch. Gen. Psychiat., 35,* 837–844.

ESCAMILLA, MICHAEL A.; SPESNY, MITZI; REUS, VICTOR I.; GALLEGOS, ALVARO; MEZA, LUIS; MOLINA, JULIO; SANDKUIJL, LODEWIJK A.; FOURNIER, EDUARDO; LEON, PEDRO E.; SMITH, LAUREN B.; and FREIMER, NELSON B. (1996). Use of linkage disequilibrium approaches to map genes for bipolar disorder in the Costa Rican population. *Am. J. Med. Genet. (Neuropsychiatric Genet.), 67,* 244–253.

FISHER, R. A. (1918). The correlation between relatives on the supposition of Mendelian inheritance. *Transactions of the Royal Society of Edinburgh, 52,* 399–433. Repr. (1966) with commentary by P. A. P. Moran and C. A. B. Smith. London: Cambridge University Press.

FODOR, STEPHEN P. A. (1977). Massively parallel genomics. *Science, 277,* 393–395.

FREIMER, N. B.; REUS, V. I.; ESCAMILLA, M.; SPESNY, M.; SMITH, L.; SERVICE, S.; GALLEGOS, A.; MEZA, L.; BATKI, S.; VINOGRADOV, S.; LEON, P.; and SANDKUIJL, L. A. (1996a). An approach to investigating linkage for bipolar disorder using large Costa Rican pedigrees. *Am. J. Med. Genet. (Neuropsychiatric Genet.), 67,* 254–263.

FREIMER, NELSON B.; REUS, VICTOR I.; ESCAMILLA, MICHAEL A.; MCINNES, L. ALISON; SPESNY, MITZI; LEON, PEDRO; SERVICE, SUSAN K.; SMITH, LAUREN B.; SILVA, SANDRA; ROJAS, EUGENIA; GALLEGOS, ALVARO; MEZA, LUIS; FOURNIER, EDUARDO; BAHARLOO, SIAMAK; BLANKENSHIP, KATHLEEN; TYLER, DAVID J.; BATKI, STEVEN; VINOGRADOV, SOPHIA; WEISSENBACH, JEAN; BARONDES, SAMUEL H.; and SANDKUIJL, LODEWIJK A. (1996b). Genetic mapping using haplotype, association and linkage methods suggests a locus for severe bipolar disorder (BPI) at 18q22-23. *Nature Genet., 12,* 436–441.

FREUD, SIGMUND. (1895). *Project for a Scientific Psychology.* In *The Standard Edition of the Complete Psychological Works of Sigmund Freud,* trans. James Strachey, I. London: Hogarth Press.

FREUD, SIGMUND. (1917). *Mourning and melancholia.* In *Collected Papers of Sigmund Freud,* trans. Joan Riviere, IV, 152–170. New York: Basic Books.

FREUD, SIGMUND. (1933). *New Introductory Lectures on Psychoanalysis,* trans. W. J. H. Sprott. New York: W. W. Norton.

FROGGAT, P., and NEVIN, N. C. (1971). The "law of ancestral heredity" and the Mendelian-ancestrian controversy in England, 1899–1906. *J. Med. Genet., 8,* 1–36.

GALTON, FRANCIS. (1865). Hereditary talent and character. *Macmillan's Magazine, 12,* 157–166, 318–327.

GALTON, FRANCIS. (1869). *Hereditary Genius: An Inquiry into Its Laws and Consequences.* London: Macmillan.

GALTON, FRANCIS. (1876). The history of twins as a criterion of the relative powers of nature and nurture. *Royal Anthropological Institute of Great Britian and Ireland Journal, 6,* 391–406.

GALTON, FRANCIS. (1889). *Natural Inheritance.* London: Macmillan.

GARDNER, RUSSELL. (1982). Mechanisms in manic–depressive disorder: an evolutionary model. *Arch. Gen. Psychiat., 39,* 1436–1441.

GARROD, ARCHIBALD E. (1902). The incidence of alkaptonuria: a study in chemical individuality. *Lancet, ii,* 1616–1620.

GARROD, ARCHIBALD E. (1909). *Inborn Errors of Metabolism*. London: Frowde; Hodder and Stoughton.

GARROD, ARCHIBALD E. (1931). *The Inborn Factors in Disease*. Oxford: Clarendon Press.

GERSHON, E. S.; HAMOVIT, J.; GUROFF, J. J.; DIBBLE, E.; LECKMAN, J. F.; SCEERY, W.; TARGUM, S. D.; NURNBERGER, J. I., JR.; GOLDIN, L. R.; and BUNNEY, W. E., JR. (1982). A family study of schizoaffective, bipolar I, bipolar II, unipolar and normal control probands. *Arch. Gen. Psychiat., 39*, 1157–1167.

GINNS, EDWARD E.; OTT, JURG; EGELAND, JANICE A.; ALLEN, CLEONA R.; FANN, CATHY, S. J.; PAULS, DAVID L.; WEISSENBACH, JEAN; CARULLI, JOHN P.; FALLS, KATHLEEN M.; KEITH, TIM P.; and PAUL, STEVEN M. (1996). A genome-wide search for chromosomal loci linked to bipolar affective disorder in the Old Order Amish. *Nature Genet., 12*, 431–435.

GOODWIN, FREDRICK K., and JAMISON, KAY REDFIELD. (1990). *Manic-Depressive Illness*. New York: Oxford University Press.

GOTTESMAN, IRVING I., and BERTELSEN, AKSEL. (1996). Legacy of German psychiatric genetics: hindsight is always 20/20. *Am. J. Med. Genet. (Neuropyschiatric Genet.), 67*, 317–322.

GRIFFITHS, ANTHONY J. F.; MILLER, JEFFREY H.; SUZUKI, DAVID T.; LEWONTIN, RICHARD C.; GELBART, WILLIAM M. (1993). *An Introduction to Genetic Analysis* (5th ed.). New York: W. H. Freeman.

GUSELLA, JAMES F.; WEXLER, NANCY S.; CONNEALLY, P. MICHAEL; NAYLOR, SUSAN L.; ANDERSON, MARY ANN; TANZI, RUDOLPH; WATKINS, PAUL C.; OTTINA, KATHLEEN; WALLACE, MARGARET R.; SAKAGUCHI, ALAN Y.; YOUNG, ANN B.; SHOULSON, IRA; BONILLO, ERNESTO; and MARTIN, JOSEPH B. (1983). A polymorphic DNA marker genetically linked to Huntington's disease. *Nature, 306*, 234–238.

HAMER, DEAN H. (1996). The heritability of happiness. *Nature Genet., 14*, 125–126.

HERSHMAN, D. JABLOW, and LIEB, JULIAN. (1988). *The Key to Genius: Manic-Depression and the Creative Life*. Buffalo, N.Y.: Prometheus Books.

HODGKINSON, STEPHEN; SHERRINGTON, ROBIN; GURLING, HUGH; MARCHBANKS, ROGER; REEDERS, STEPHEN; MALLET, JACQUES; MCINNIS, MELVIN; PETURSON, HANNES; and BRYNJOLFSSON, JON. (1987). Molecular genetic evidence for heterogeneity in manic depression. *Nature, 325*, 805–806.

HUNTINGTON, GEORGE W. (1872). On chorea. *Med. Surgical Reporter, 26,* 317–321.

HUNTINGTON'S DISEASE COLLABORATIVE RESEARCH GROUP. (1993). A novel gene containing a trinucleotide repeat that is expanded and unstable on Huntington's disease chromosomes. *Cell, 72,* 971–983.

HUGGIN, MARLENE; BLOCH, MAURICE; KANANI, SHELIN; QUARRELL, OLIVER W. J.; THEILMAN, JANE; HEDRICK, AMY; DICKENS, BERNARD; LYNCH, ABYANN; and HAYDEN, MICHAEL. (1990). Ethical and legal dilemmas arising during predictive testing for adult-onset disease: the experience of Huntington's disease. *Am. J. Hum. Genet., 47,* 4–12.

HUXLEY, JULIAN; MAYR, ERNST; OSMOND, HUMPHRY; and HOFFER, ABRAM. (1964). Schizophrenia as a genetic morphism. *Nature, 204,* 220–221.

INGRAM, V. M. (1957). Gene mutations in human hemoglobin: the chemical difference between normal and sickle cell hemoglobin. *Nature, 180,* 326–328.

JAMISON, KAY REDFIELD. (1993). *Touched with Fire: Manic-Depressive Illness and the Artistic Temperament.* New York: Free Press.

JAMISON, KAY REDFIELD. (1995). *An Unquiet Mind.* New York: Alfred A. Knopf.

JENKINS, JR., HOLMAN W. (1996, October 8). The latest management craze: crazy management. *The Wall Street Journal.*

JORDAN, BARRY D.; RELKIN, NORMAN R.; RAVDIN, LISA D.; JACOBS, ALAN R.; BENNETT, ALEXANDRE; and GANDY, SAM. (1997). Apolipoprotein E4 associated with chronic traumatic brain injury in boxing. *JAMA, 278,* 136–140.

JUDSON, HORACE FREELAND. (1979). *The Eighth Day of Creation.* New York: Simon and Schuster.

KAN, YUET WAI, and DOZY, ANDRÉE M. (1978). Polymorphism of DNA sequence adjacent to human beta-globin structural gene: relationship to sickle mutation. *Proc. Natl. Acad. Sci. USA, 75,* 5631–5635.

KAREL, RICHARD B. (1994, November 18). Szasz settles suit for $650,000. *Psychiat. News, 29,* 22.

KELSOE, JOHN R.; GINNS, EDWARD I.; EGELAND, JANICE A.; GERHARD, DANIELA S.; GOLDSTEIN, ALISA M.; BALE, SHERRI J.; PAULS, DAVID L.; LONG, ROBERT T.; KIDD, KENNETH K.; CONTE, GIOVANNI; HOUSMAN, DAVID E.; and PAUL, STEVEN M. (1989). Re-evaluation of the linkage relationship between chromosome 11p loci and the gene for bipolar affective disorder in the Old Order Amish. *Nature, 342,* 238–243.

KENDLER, KENNETH S. (1995). Genetic epidemiology in psychiatry: taking both genes and environment seriously. *Arch. Gen. Psychiat., 52,* 895–899.

KEVLES, DANIEL J. (1986). *In the Name of Eugenics.* Berkeley: University of California Press.

KEVLES, DANIEL J., and HOOD, LEROY, eds. (1992). *The Code of Codes: Scientific and Social Issues in the Human Genome Project.* Cambridge, MA: Harvard University Press.

KIRK, STUART A., and KUTCHINS, HERB. (1992). *The Selling of DSM: The Rhetoric of Science in Psychiatry.* New York: Aldine De Gruyter.

KITCHER, PHILIP. (1996). *The Lives to Come: The Genetic Revolution and Human Possibilities.* New York: Simon and Schuster.

KOVACS, MARIA; DEVLIN, BERNIE; POLLOCK, MYRNA; RICHARDS, CHERYL; and PROTAP, MUKERJI. (1997). A controlled family history study of child-hood-onset depressive disorder. *Arch. Gen. Psychiat., 54,* 613–623.

KRAEPELIN, EMIL. (1915). *Manic-Depressive Insanity and Paranoia,* trans. (1921) R. Mary Barclay and George M. Robertson. Edinburgh: E. and S. Livingstone.

KRAMER, PETER D. (1993). *Listening to Prozac.* New York: Viking.

KRUGLYAK, LEONID. (1997). The use of a genetic map of biallelic markers in linkage studies. *Nature Genet., 17,* 21–24.

KUNUGI, H.; HATTORI, M.; KATO, T.; TATSUMI, M.; SAKAI, T.; SASAKI, T.; HIROSE, T.; and NANKO, S. (1997). Serotonin transporter gene polymor-phisms: ethnic difference and possible association with bipolar affective dis-order. *Molecular Psychiat., 2,* 457–462.

LANDER, ERIC S. (1996). The new genomics: global views of biology. *Science, 274,* 536–539.

LANDER, ERIC S., and KRUGLYAK, LEONID. (1995). Genetic dissection of complex traits: guidelines for interpreting and reporting linkage results. *Nature Genet., 11,* 241–247.

LANDER, ERIC S., and SCHORK, NICHOLAS J. (1994). Genetic dissection of complex traits. *Science, 265,* 2037–2048.

LENDON, CONNIE L.; ASHALL, FRANK; and GOATE, ALISON M. (1997). Exploring the etiology of Alzheimer disease using molecular genetics. *JAMA, 277,* 825–831.

LEON, PEDRO; RAVENTOS, HENRIETTE; LYNCH, ERIC; MORROW, JAN; and KING, MARY-CLAIRE. (1992). The gene for an inherited form of deafness maps to chromosome 5q31. *Proc. Natl. Acad. Sci. USA, 89,* 5181–5184.

LEONHARD, KARL. (1957). *The Classification of Endogenous Psychoses,* ed. Eli Robins, trans. (1997) Russell Berman. New York: Irvington Publishers.

LIFTON, RICHARD P. (1996). Molecular genetics of human blood pressure variation. *Science, 272,* 676–680.

LIFTON, ROBERT JAY. (1986). *The Nazi Doctors: Medical Killing and the Psychology of Genocide.* New York: Basic Books.

LYKKEN, DAVID, and TELLEGEN, AUKE. (1996). Happiness is a stochastic phenomenon. *Psychological Science, 7,* 186–189.

LYKKEN, D. T.; MCGUE, M.; TELLEGEN, A.; and BOUCHARD, JR., T. J. (1992). Emergenesis: genetic traits that may not run in families. *Am. Psychologist, 12,* 1565–1577.

LYNCH, ERIC D.; LEE, MING K.; MORROW, JAN E.; WELCSH, PIRI L.; LÉON, PEDRO E.; and KING, MARY-CLAIRE. (1997). Nonsyndromic deafness DFNA1 associated with mutation of a human homolog of the *Drosophila* gene *diaphenous. Science, 278,* 1315–1318.

MACALPINE, IDA, and HUNTER, RICHARD. (1966). The "insanity" of King George III: a classic case of porphyria. *Br. Med. J., 1,* 65–71.

MACALPINE, IDA, and HUNTER, RICHARD. (1969). *George III and the Mad Business.* London: Allen Lane/Penguin Press.

MANJI, HUSSEINI K.; POTTER, WILLIAM Z.; and LENOX, ROBERT H. (1995). Signal transduction pathways: molecular targets for lithium's actions. *Arch. Gen. Psychiat., 52,* 531–543.

MAROTEAUX, PIERRE, and LAMY, MAURICE. (1965). The malady of Toulouse-Lautrec. *JAMA, 191,* 111–113.

MAYEUX, R.; OTTMAN, R.; MAESTRE, G.; NGAI, B. S.; TANG, M.-X.; GINSBERG, H.; CHUN, M.; TYCKO, B.; and SHELANSKI, M. (1995). Synergistic effects of traumatic head injury and apolipoprotein-E4 in patients with Alzheimer's disease. *Neurology, 45,* 555–557.

MAYNARD SMITH, JOHN. (1989). *Evolutionary Genetics.* Oxford: Oxford University Press.

MAYR, E. (1961). Cause and effect in biology. *Science, 134,* 1501–1506.

MCGUFFIN, PETER, and KATZ, RANDY. Nature, nuture, and affective disorder. In J. F. W. Deakin, ed., *The Biology of Depression,* 26–51. London: Gaskell.

MCGUFFIN, PETER; OWEN, MICHAEL J.; O'DONOVAN, MICHAEL C.; THAPAR, ANITA; and GOTTESMAN, IRVING I. (1994). *Seminars in Psychiatric Genetics.* London: Gaskell.

McGuire, M. T.; Marks, I.; Nesse, R. M.; and Troisi, A. (1992). Evolutionary biology: a basic science for psychiatry? *Acta Psychiat. Scand.*, *86*, 89–96.

McInnes, L. Alison; Escamilla, Michael A.; Service, Susan K.; Reus, Victor I.; Leon, Pedro; Silva, Sandra; Rojas, Eugenia; Spesny, Mitzi; Baharloo, Siamak; Blankenship, Kathleen; Peterson, Amy; Tyler, David; Shimayoshi, Norito; Tobey, Christa; Batki, Steven; Vinogradov, Sophia; Meza, Luis; Gallegos, Alvaro; Fournier, Eduardo; Smith, Lauren B.; Barondes, Samuel H.; Sandkuijl, Lodewijk A.; and Freimer, Nelson B. (1996). A complete genome screen for genes predisposing to severe bipolar disorder in two Costa Rican pedigrees. *Proc. Natl. Acad. Sci. USA*, *93*, 13060–13065.

McKusick, Victor A. (1978). *Medical Genetic Studies of the Amish.* Baltimore: Johns Hopkins University Press.

McKusick, Victor A. (1991). The defect in Marfan syndrome. *Nature*, *352*, 279–281.

Mendel, Gregor. (1866). Experiments on plant hybrids, trans. Eva R. Sherwood. In Curt Stern and Eva R. Sherwood, eds., *The Origin of Genetics*, (1966), San Francisco: W. H. Freeman.

Menninger, Karl. (1963). *The Vital Balance.* New York: Viking.

Muller-Hill, Benno. (1988). *Murderous Science.* Oxford: Oxford University Press.

Nesse, Randolph M., and Williams, George C. (1995). *Why We Get Sick: The New Science of Darwinian Medicine.* New York: Times Books.

Nirenberg, Marshall W.; Matthaei, J. Heinrich; Jones, Oliver W.; Martin, Robert G.; and Barondes, Samuel H. (1963). Approximation of genetic code via cell-free protein synthesis directed by template RNA. *Federation Proc.*, *22*, 55–61.

Nurnberger, Jr., John I.; Blehar, Mary C.; Kaufmann, Charles A.; York-Cooler, Carolyn; Simpson, Sylvia G.; Harkavy-Friedman, Jill; Severe, Joanne B.; Malaspina, Dolores; Reich, Theodore; and Collaborators from the NIMH Genetics Initiative. (1994). Diagnostic interview for genetic studies. *Arch. Gen. Psychiat.*, *51*, 849–859.

Ogilvie, Alan D.; Battersby, Sharon; Bubb, Vivien J.; Fink, George; Harmar, Anthony J.; Goodwin, Guy M.; and Smith, C. A. Dale. (1996). Polymorphism in serotonin transporter gene associated with susceptibility to major depression. *Lancet*, *347*, 731–733.

PAULING, L.; ITANO, H. A.; SINGER, S. J.; and WELLS, I. C. (1949). Sickle cell anemia: a molecular disease. *Science, 110,* 543–548.

PENROSE, L. S. (1953). The genetical background of common diseases. *Acta Genet., 4,* 257–265.

PEARSON, VERONICA. (1995). Population policy and eugenics in China. *Br. J. Psychiat., 167,* 1–4.

PERICAK-VANCE, M. A.; BASS, M. P.; YAMAOKA, L. H.; GASKELL, P. C.; SCOTT, W. K.; TERWEDOW, H. A.; MENOLD, M. M.; CONNEALLY, P. M.; SMALL, G. W.; VANCE, J. M.; SAUNDERS, A. M.; ROSES, A. D.; and HAINES, J. L. (1997). Complete genome screen in late-onset familial Alzheimer disease. Evidence for a new locus on chromosome 12. *JAMA, 278,* 1237–1241.

PERICAK-VANCE, M. A.; BEBOUT, J. L.; GASKELL, JR., P. C.; YAMAOKA, L. H.; HUNG, W.-Y.; ALBERTS, M. J.; WALKER, A. P.; BARTLETT, R. J.; HAYNES, C. A.; WELSH, K. A.; EARL, N. L.; HEYMAN, A.; CLARK, C. M.; and ROSES, A. D. (1991). Linkage studies in familial Alzheimer disease: evidence for chromosome 19 linkage. *Am. J. Hum. Genet., 48,* 1034–1050.

PLOMIN, R.; CHIPUER, H. M.; and NIEDERHISER, J. M. (1994). Behavioral genetic evidence for the importance of nonshared environment. In E. M. Hetherington, D. Reiss, and R. Plomin, eds., *Separate Social Worlds of Siblings: Impact of Nonshared Environment on Development,* 1–31. Hillsdale, N.J.: Lawrence Erlbaum.

PLOMIN, ROBERT, and DANIELS, DENISE. (1987). Why are children in the same family so different from each other? *Behav. Brain Sci., 10,* 1–60.

PLOMIN, ROBERT; DEFRIES, JOHN C.; McCLEARN, GERALD E.; and RUTTER, MICHAEL. (1997). *Behavioral Genetics* (3d ed.). New York: W. H. Freeman.

PLOMIN, ROBERT; OWEN, MICHAEL J.; and McGUFFIN, PETER. (1994). The genetic basis of complex human behaviors. *Science, 264,* 1733–1739.

POST, STEPHEN G.; WHITEHOUSE, PETER J.; BINSTOCK, ROBERT H.; BIRD, THOMAS D.; ECKERT, SHAREN K.; FARRER, LINDSAY A.; FLECK, LEONARD M.; GAINES, ATWOOD D.; JUENGST, ERIC T.; KARLINSKY, HARRY; MILES, STEVEN; MURRAY, THOMAS H.; QUAID, KIMBERLY A.; RELKIN, NORMAN R.; ROSES, ALLEN D.; ST. GEORGE-HYSLOP, P. H.; SACHS, GREG A.; STEINBOCK, BONNIE; TRUSCHKE, EDWARD F.; and ZINN, ARTHUR B. (1997). The clinical introduction of genetic testing for Alzheimer disease: an ethical perspective. *JAMA, 27,* 832–836.

RICHARDS, RUTH; KINNEY, DENNIS K.; LUNDE, INGE; BENET, MARIA; and MERZEL, ANN P. C. (1988). Creativity in manic-depressives, cyclothymes, their normal relatives, and control subjects. *J. Abnormal Psychology, 97,* 281–288.

RISCH, NEIL. (1990a). Genetic linkage and complex diseases, with special reference to psychiatric disorders. *Genet. Epidemiology, 7,* 3–16.

RISCH, NEIL. (1990b). Linkage strategies for genetically complex traits. I. Multilocus models. *Am. J. Hum. Genet., 46,* 222–228.

RISCH, NEIL, and BOTSTEIN, DAVID. (1996). A manic depressive history. *Nature Genet., 12,* 351–353.

RISCH, NEIL, and MERIKANGAS, KATHLEEN. (1996). The future of genetic studies of complex human diseases. *Science, 273,* 1516–1517.

ROBERTSON, MIRANDA. (1989). False start on manic depression. *Nature, 342,* 222.

ROBERTSON, MIRANDA. (1987). Molecular genetics of the mind. *Nature, 325,* 755.

ROSES, ALLEN D. (1995). Apolipoprotein E genotyping in the differential diagnosis, not prediction, of Alzheimer's disease. *Annals of Neurology, 38,* 6–14.

ROSES, ALLEN D. (1996). Apolipoprotein E alleles as risk factors in Alzheimer's disease. *Ann. Rev. Med., 47,* 387–400.

ROSES, ALLEN D. (1997). A model for susceptibility polymorphisms for complex diseases: apolipoprotein E and Alzheimer disease. *Neurogenetics, 1,* 3–11.

RUWENDE, C.; KHOO, S. C.; SNOW, R. W.; YATES, S. N. R.; KWIATKOWSKI, D.; GUPTA, S.; WARN, P.; ALSOPP, C. E. M.; GILBERT, S. C.; PESCH, N.; NEWBOLD, C. I.; GREENWOOD, B. M.; MARSH, K.; and HILL, A. V. S. (1995). Natural selection of hemi- and heterozygotes for G6PD deficiency in Africa by resistance to severe malaria. *Nature, 376,* 246–249.

SCHATZBERG, ALAN F., and NEMEROFF, CHARLES B., eds. (1995). *The American Psychiatric Press Textbook of Psychopharmacology.* Washington, D.C.: American Psychiatric Press.

SCRIVER, CHARLES R. (1996). Alkaptonuria: such a long journey. *Nature Genet., 14,* 5–6.

SCRIVER, CHARLES R.; BEAUDET, ARTHUR L.; SLY, WILLIAM S.; and VALLE, DAVID. (1995). *The Metabolic and Molecular Bases of Inherited Disease* (7th ed.). New York: McGraw-Hill.

SELKOE, DENNIS J. (1997). Alzheimer's disease: genotypes, phenotype, and treatments. *Science, 275,* 630–631.

SHINE, IAN, and WROBEL, SYLVIA. (1976). *Thomas Hunt Morgan: Pioneer of Genetics.* Lexington: University of Kentucky Press.

SHORTER, EDWARD. (1996). *A History of Psychiatry.* New York: John Wiley.

SIMMONS, JOHN. (1996). *The Scientific 100: A Ranking of the Most Influential Scientists, Past and Present.* Secaucus, N.J.: Carol Publishing Group.

SINERVO, B., and LIVELY, C. M. (1996). The rock-paper-scissors game and the evolution of alternative male strategies. *Nature, 380,* 240–243.

SMITH, LAUREN B.; SAPERS, BENJAMIN; REUS, VICTOR I.; and FREIMER, NELSON B. (1996). Attitudes toward bipolar disorder and predictive genetic testing among patients and providers. *J. Med. Genet., 33,* 544–549.

SNYDER, SOLOMON H. (1996). *Drugs and the Brain.* New York: Scientific American Library.

SPENCE, M. ANNE; FLODMAN, PAMELA L.; SADOVNICK, A. DESSA; BAILEY-WILSON, JOAN E.; AMELI, HOSSEIN; and REMICK, RONALD A. (1995). Bipolar disorder: evidence for a major locus. *Am. J. Med. Genet. (Neuropsychiatric Genet.), 60,* 370–376.

STINE, O. COLIN; XU, JIANFENG; KOSKELA, REBECCA; MCMAHON, FRANCIS I.; GSCHWEND, MICHELE; FRIDDLE, CARL; CLARK, CHRIS D.; MCINNIS, MELVIN G.; SIMPSON, SILVIA G.; BRESCHEL, THERESA S.; VISHIO, EVA; RISKIN, KELLY; FEILOTTER, HARRIET; CHEN, EUGENE; SHEN, SUSAN; FOLSTEIN, SUSAN; MEYERS, DEBORAH; BOTSTEIN, DAVID; MARR, THOMAS G.; and DEPAULO, J. RAYMOND. (1995). Evidence for linkage of bipolar disorder to chromosome 18 with a parent-of-origin effect. *Am. J. Hum. Genet., 57,* 1384–1394.

STONE, ALAN A. (1997, January). Where will psychoanalysis survive? What remains of Freudianism when its scientific center crumbles? *Harvard Magazine,* 35–39.

STRACHAN, TOM, and READ, ANDREW P. (1996). *Human Molecular Genetics.* Oxford: Bios Scientific Publishers.

STURTEVANT, A. H. (1913). The linear arrangement of six sex-linked factors in Drosophila, as shown by their mode of association. *J. Experimental Zoology, 14,* 43–59.

STURTEVANT, A. H. (1940). A new inherited character in man. *Proc. Natl. Acad. Sci. USA, 26,* 100–102.

STURTEVANT, A. H. (1965). *A History of Genetics.* New York: Harper and Row.

SULLOWAY, FRANK J. (1979). *Freud, Biologist of the Mind.* New York: Basic Books.

SZASZ, THOMAS S. (1961). *The Myth of Mental Illness.* New York: Harper and Row.

TELLEGEN, A.; LYKKEN, D. T.; BOUCHARD, T. J.; WILCOX, K.; SEGAL, N.; and RICH, S. (1988). Personality similarity in twins reared apart and together. *J. Pers. Soc. Psychol., 54,* 1031–1039.

TODD, JOHN A. (1996). Transcribing diabetes. *Nature, 384,* 407–408.

TSUANG, MING T., and FARAONE, STEPHEN T. (1990). *The Genetics of Mood Disorders.* Baltimore: Johns Hopkins University Press.

TURECKI, GUSTAVO; ROULEAU, GUY A.; MARI, JAIR; JOOBER, RIDHA; and MORGAN, KENNETH. (1997). Lack of association between bipolar disorder and tyrosine hydroxylase: a meta-analysis. *Am. J. Med. Genet. (Neuropsychiatric Genet.), 74,* 348–352.

VOGEL, FRIEDRICH, and MOTULSKY, ARNO G. (1997). *Human Genetics* (3d ed.). Berlin: Springer-Verlag.

WATSON, J. D., and CRICK, F. H. C. (1953). Molecular structure of nucleic acids: a structure for deoxyribose nucleic acid. *Nature, 171,* 737–738.

WATSON, JAMES D.; GILMAN, MICHAEL; WITKOWSKI, JAN; and ZOLLER, MARK. (1992). *Recombinant DNA* (2d ed.). New York: Scientific American Books.

WEBER, JAMES L., and MAY, PAULA E. (1989). Abundant class of human DNA polymorphisms which can be typed using the polymerase chain reaction. *Am. J. Hum. Genet., 44,* 388–396.

WEISS, KENNETH M. (1995). *Genetic Variation and Human Disease: Principles and Evolutionary Approaches.* Cambridge: Cambridge University Press.

WEISSMAN, MYRNA M.; BLAND, ROGER C.; CANINO, GLORISA J.; FARA-VELLI, CARLO; GREENWALD, STEVEN; HWU, HAI-GWO; JOYCE, PETER R.; KARAM, ELIE G.; LEE, CHUNG-KYOON; LELLOUCH, JOSEPH; LEPINE, JEAN-PIERRE; NEWMAN, STEPHEN C.; RUBIO-STIPEC, MARITZA; WELLS, J. ELIZABETH; WICKRAMARANTE, PRIYA J.; WITTCHEN, HANS-ULRICH; and YEH, ENG-KUNG. (1996). Cross-national epidemiology of major depression and bipolar disorder. *JAMA, 276,* 293–299.

WEISSMAN, MYRNA M.; WARNER, VIRGINIA; WICKRAMARANTE, PRIYA; and PRUSOFF, BRIGITTE A. (1988). Early-onset major depression in parents and their children. *J. Affective Disorders, 15,* 269–277.

WEISSMAN, MYRNA M.; WICKRAMARANTE, PRIYA; MERIKANGAS, KATHLEEN R.; LECKMAN, JAMES F.; PRUSOFF, BRIGITTE A.; CARUSO, KEITH A.; KIDD, KENNETH K.; and GAMMON, DAVIS. (1984). Onset of major depression in early adulthood: increased familial loading and specificity. *Arch. Gen. Psychiat., 41,* 1136–1143.

WERTZ, DOROTHY C.; FLETCHER, JOHN C.; and MULVIHIL, JOHN J. (1990). Medical geneticists confront ethical dilemmas: cross-cultural comparisons among 18 nations. *Am. J. Hum. Genet., 46,* 1200–1213.

WEXLER, ALICE. (1995). *Mapping Fate.* New York: Times Books.

WHYBROW, PETER C. (1997). *A Mood Apart.* New York: Basic Books.

WINOKUR, GEORGE; CORYELL, WILLIAM; KELLER, MARTIN; ENDICOTT, JEAN; and LEON, ANDREW. (1995). A family study of manic-depressive (bipolar I) disease: is it a distinct illness separable from primary unipolar depression? *Arch. Gen. Psychiat., 52,* 367–373.

ZILBOORG, GREGORY. (1941). *A History of Medical Psychology.* New York: W. W. Norton.

ACKNOWLEDGMENTS

This book was made possible by longstanding support from two great public institutions. The University of California has provided me with an exceptionally generative environment for almost three decades, first at its San Diego campus, and since 1986, at its San Francisco campus (UCSF). The National Institutes of Health has funded my research since 1960, most recently with grant MH 47563, "Genetics of Bipolar Disorder."

I have also been blessed with many wonderful colleagues. I thank David Housman for helping me realize that it was becoming feasible in the early 1980s to hunt for genes that play a role in mental illness, and David Cox for joining me several years later to found the Neurogenetics Laboratory at UCSF. I thank Nelson Freimer, for establishing the Costa Rican project on the genetics of manic-depressive illness in which I participate, as well as other key collaborators: Michael Escamilla, Pedro Léon, Alison McInnes, Victor Reus, Lodewijk Sandkuijl, and Mitzi Spesny.

Several colleagues read an earlier draft of this book and provided extremely helpful detailed comments: Chandler Burr, Nelson Freimer, Ira Herskowitz, Steve Hyman, Victor Reus, and Martin Zatz. Several others provided important suggestions: Charlie Epstein, Kay Redfield Jamison, Tom Kornberg, David Kupfer, Barry Sinervo, Ming Tsuang, Myrna Weissman, Peter Whybrow, and Andrea Zanko. While writing this book I had the privilege of serving as chair of the National Institute of Mental Health's Workgroup on Genetics, and was influenced greatly by discussions with its members: Aravinda Chakravarti, Mary-Claire King, Eric Lander, Bob Nussbaum, Ted Reich, Joe Takahashi, and Steve Warren.

I benefited immensely from the perceptive insights and warm camaraderie of Jonathan Cobb, Senior Editor at W. H. Freeman. I thank Nancy Brooks for clarifying and enlivening my prose, Hilary Hinzmann for discerning comments, and Maria Epes for help with art and design. I thank my literary agent, Katinka Matson, President of Brockman Inc., for valuable advice.

I am particularly appreciative of the generosity and the enduring friendship of Jeanne and Sandy Robertson, who endowed the Chair in Neurobiology and Psychiatry at UCSF which I occupy. Most of all I thank Louann Brizendine, Elizabeth Barondes, and Jessica Barondes, who have each made many helpful suggestions on drafts of this book, and who bring joy to my life in so many ways.

INDEX

READ MORE IN PENGUIN

READ MORE IN PENGUIN

SCIENCE AND MATHEMATICS

Six Easy Pieces Richard P. Feynman

Drawn from his celebrated and landmark text *Lectures on Physics*, this collection of essays introduces the essentials of physics to the general reader. 'If one book was all that could be passed on to the next generation of scientists it would undoubtedly have to be *Six Easy Pieces*' John Gribbin, *New Scientist*

A Mathematician Reads the Newspapers John Allen Paulos

In this book, John Allen Paulos continues his liberating campaign against mathematical illiteracy. 'Mathematics is all around you. And it's a great defence against the sharks, cowboys and liars who want your vote, your money or your life' Ian Stewart

Dinosaur in a Haystack Stephen Jay Gould

'Today we have many outstanding science writers ... but, whether he is writing about pandas or Jurassic Park, none grabs you so powerfully and personally as Stephen Jay Gould ... he is not merely a pleasure but an education and a chronicler of the times' *Observer*

Does God Play Dice? Ian Stewart

As Ian Stewart shows in this stimulating and accessible account, the key to this unpredictable world can be found in the concept of chaos, one of the most exciting breakthroughs in recent decades. 'A fine introduction to a complex subject' *Daily Telegraph*

About Time Paul Davies

'With his usual clarity and flair, Davies argues that time in the twentieth century is Einstein's time and sets out on a fascinating discussion of why Einstein's can't be the last word on the subject' *Independent on Sunday*

READ MORE IN PENGUIN

SCIENCE AND MATHEMATICS

In Search of SUSY John Gribbin

Many physicists believe that we are on the verge of developing a complete 'theory of everything' which can reduce the four basic forces of nature – gravity, electromagnetism, the strong and weak nuclear forces – to a single superforce. At its heart is the principle of super-symmetry (SUSY).

Fermat's Last Theorem Amir D. Aczel

Here, weaving together history and science, Amir D. Aczel offers a thrilling, step-by-step account of the search for the mathematicians' Holy Grail. 'Mr Aczel has written a tale of buried treasure . . . This is a captivating volume' *The New York Times*

Insanely Great Steven Levy

It was Apple's co-founder Steve Jobs who referred to the Mac as 'insanely great'. He was absolutely right: the machine that revolu-tionized the world of personal computing was and is great – yet the machinations behind its inception were nothing short of insane. 'A delightful and timely book' *The New York Times Book Review*

The Artful Universe John D. Barrow

This thought-provoking investigation illustrates some unexpected links between art and science. 'Full of good things . . . In what is probably the most novel part of the book, Barrow analyses music from a mathematical perspective . . . an excellent writer' *New Scientist*

The Jungles of Randomness Ivars Peterson

Taking us on a fascinating journey into the ambiguities and un-certainties of randomness, Ivars Peterson explores the complex interplay of order and disorder, giving us a new understanding of nature and human activity.

READ MORE IN PENGUIN

SCIENCE AND MATHEMATICS

The Fabric of Reality   David Deutsch

'Reading this book might just change your life ... The theory of everything, quantum mechanics, virtual reality, evolution, the significance of life, time travel, the end of our Universe: all these and much else find their place ... this is an awesome book' *New Scientist*

Mathematics: The New Golden Age   Keith Devlin

In the computerized world of today mathematics has an impact on almost every aspect of our lives, yet most people believe they cannot hope to understand or enjoy the subject. This comprehensive survey sets out to show just how mistaken they are and brilliantly captures the essential richness of mathematics' 'new golden age'.

Climbing Mount Improbable   Richard Dawkins

'Mount Improbable ... is Dawkins's metaphor for natural selection: its peaks standing for evolution's most complex achievements ... exhilarating – a perfect, elegant riposte to a great deal of fuzzy thinking about natural selection and evolution' *Observer*. 'Dazzling' David Attenborough

Brainchildren   Daniel C. Dennett

Thinking about thinking can be a baffling business. Investigations into the nature of the mind – how it works, why it works, its very existence – can seem convoluted to the point of fruitlessness. Daniel C. Dennett has provided an eloquent and often witty guide through some of the mental and moral mazes that surround these areas of thought.

Fluid Concepts and Creative Analogies   Douglas Hofstadter

'This exhilarating book is the most important on AI for the thoughtful general reader in years and ample proof that reports of the death of artificial intelligence have been greatly exaggerated' *The New York Times Book Review*

READ MORE IN PENGUIN

POPULAR SCIENCE

In Search of Nature   Edward O. Wilson

A collection of essays of 'elegance, lucidity and breadth' *Independent*. 'A graceful, eloquent, playful and wise introduction to many of the subjects he has studied during his long and distinguished career in science' *The New York Times*

Clone   Gina Kolata

'A thoughtful, engaging, interpretive and intelligent account ... I highly recommend it to all those with an interest in ... the new developments in cloning' *New Scientist*. 'Superb but unsettling' J. G. Ballard, *Sunday Times*

The Feminization of Nature   Deborah Cadbury

Scientists around the world are uncovering alarming facts. There is strong evidence that sperm counts have fallen dramatically. Testicular and prostate cancer are on the increase. Different species are showing signs of 'feminization' or even 'changing sex'. 'Grips you from page one ... it reads like a Michael Crichton thriller' John Gribbin

Richard Feynman: A Life in Science   John Gribbin and Mary Gribbin

'Richard Feynman (1918–88) was to the second half of the century what Einstein was to the first: the perfect example of scientific genius' *Independent*. 'One of the most influential and best-loved physicists of his generation ... This biography is both compelling and highly readable' *Mail on Sunday*

***T. rex* and the Crater of Doom**   Walter Alvarez

Walter Alvarez unfolds the quest for the answer to one of science's greatest mysteries – the cataclysmic impact on Earth which brought about the extinction of the dinosaurs. 'A scientific detective story par excellence, told with charm and candour' Niles Eldredge

READ MORE IN PENGUIN

POPULAR SCIENCE

How the Mind Works Steven Pinker

'Presented with extraordinary lucidity, cogency and panache ... Powerful and gripping ... To have read [the book] is to have consulted a first draft of the structural plan of the human psyche ... a glittering *tour de force*' *Spectator*. 'Witty, lucid and ultimately enthralling' *Observer*

At Home in the Universe Stuart Kauffman

Stuart Kauffman brilliantly weaves together the excitement of intellectual discovery and a fertile mix of insights to give the general reader a fascinating look at this new science – the science of complexity – and at the forces for order that lie at the edge of chaos. 'Kauffman shares his discovery with us, with lucidity, wit and cogent argument, and we see his vision ... He is a pioneer' Roger Lewin

Stephen Hawking: A Life in Science
Michael White and John Gribbin

'A gripping account of a physicist whose speculations could prove as revolutionary as those of Albert Einstein ... Its combination of erudition, warmth, robustness and wit is entirely appropriate to their subject' *New Statesman & Society*. 'Well-nigh unputdownable' *The Times Educational Supplement*

Voyage of the *Beagle* Charles Darwin

The five-year voyage of the *Beagle* set in motion the intellectual currents that culminated in the publication of *The Origin of Species*. His journal, reprinted here in a shortened version, is vivid and immediate, showing us a naturalist making patient observations, above all in geology. The editors have provided an excellent introduction and notes for this edition, which also contains maps and appendices.